Working with Nature

Working with Nature

Resource Management for Sustainability

Carl F. Jordan

*Institute of Ecology
University of Georgia
Athens, Georgia, USA*

 harwood academic publishers

Australia • Canada • China • France • Germany • India •
Japan • Luxembourg • Malaysia • The Netherlands •
Russia • Singapore • Switzerland

Copyright © 1998 OPA (Overseas Publishers Association) N.V. Published by license under the Harwood Academic Publishers imprint, part of The Gordon and Breach Publishing Group.

All rights reserved.

No part of this book may be reproduced or utilized in any form or by any means, electronic or mechanical, including photocopying and recording, or by any information storage or retrieval system, without permission in writing from the publisher. Printed in Singapore.

Amsteldijk 166
1st Floor
1079 LH Amsterdam
The Netherlands

British Library Cataloguing in Publication Data

Jordan, Carl F.
  Working with nature : resource management for sustainability
  1.Sustainable development
  I.Title
  333.7'17

ISBN 90-5702-542-6

Contents

Preface vii

INTRODUCTION 1

1 THE RATCHET EFFECT 5
   The Mechanical Round 5
   The Chemical Round 9
   The Genetic Round 14
   The Tightening Ratchet 19

2 THE PROBLEM OF SCALE 21
   Interaction with Technology 21
   Meat and Dairy Production 22
   Scale and Forestry 24
   Irrigation and Scale 25
   Scale and Loss of Species Diversity 25

3 AN ALTERNATIVE PHILOSOPHY 27
   Working with Nature 28
   Economic Feasibility 28
   Conclusion 30

4 THE BOTTOMS-UP APPROACH 31

5 PLANT–PLANT INTERACTIONS 35
   Negative Interactions 35
   Positive Interactions 40
   Intercropping in Agriculture 42
   Mixed Species Forest Plantations 45
   Management for Facilitation and Complementarity 49

6 INTERACTIONS IN THE SOIL 51
   Functions of Soil Organisms 51
   Changes in the Soil System Following Cultivation 58
   How We Compensate for the Loss of Nature's Services ... 61
   Conclusion 67

vi / Contents

7 PEST AND DISEASE INTERACTIONS 69
   Pests 69
   Disease 71
   Integrated Pest Management 73
   Economic Analysis 80
   Environmental Disturbance, and Pests and Disease 81
   Conclusion 83

8 INTERACTIONS DURING SUCCESSION 85
   Factors Influencing Succession 85
   Trends During Succession 87
   Fighting Succession 91
   Working with Succession 94
   Conclusion 96

9 THE TOP-DOWN APPROACH 99
   An Alternative Approach 99
   Energy and Power Output 103
   Guiding Principle for Resource Management 111

10 CASE STUDIES OF RESOURCE MANAGEMENT 113
   Transition from Sustainability to Unsustainability 114
   Transition from Unsustainability to Sustainability 125
   Factors Causing Transitions 132

11 CONCLUSION 135

Literature Cited 139

Subject Index 159

Author Index 165

PREFACE

Increasing the yield of farms, forests, and rangelands has been an important goal for resource managers throughout the world. For centuries, farmers, foresters, and ranchers have experimented with new technology and new ideas in an effort to bolster the productivity of crops, forests, and grasslands. Until the mid-nineteenth century, most advances depended on empirical observation: A farmer tried something new and if it increased output or profit, he repeated it. An important change in the approach to increasing resource yield occurred in the United States in 1862, however, with the passage of the Morrill Land-Grant College Act. This Act established colleges specifically dedicated to problems of applied management and, in so doing, elevated the quest for increased productivity from a trial-and-error approach to a systematic approach based on scientific method.

This application of science to increase production of crops, livestock, and trees has been immensely successful. Sufficient food and fiber for entire populations can now be produced by a small proportion of society, thereby freeing people from the necessity of farming for subsistence, and allowing them to pursue other goals that epitomize civilization — commerce, engineering, arts, and literature, to name a few.

In recent decades, however, increasing ecological side effects of technological agriculture, forestry, and range management have prompted a citizens' call for environmental quality as an end equally important as high yield. Scientists have responded by devising new technologies that ameliorate problems caused by the narrow focus upon productivity; however, becoming dependent upon technology to solve environmental problems has its own disadvantages. One such disadvantage is that our technology is not sophisticated enough to carry out all the life-support functions performed by nature. A case in point is Biosphere 2, the $200-million closed-environment facility intended to create a self-sustaining habitat for humans, plants, and animals. Biosphere 2 encountered numerous problems and surprises, even though almost unlimited energy and technology were available for its support. The facility has proven that it is not yet possible to engineer a system that can support just eight human beings with adequate food, water, and air for two years (Cohen and Tilman 1996).

Even if a system to control nature were theoretically perfect, technology and the humans that control technology are fallible. Because increases in production concentrate resources and their by-products, a breakdown in technology or a human mistake has increasingly severe consequences. Another drawback of technological control is that it requires energy subsidies to maintain control. Petroleum is currently the base for sustaining agriculture, forestry, and livestock; a sharp increase in its price would cause serious disruptions in meeting the world's resource needs.

Our preoccupation with technology to solve resource management problems has led us to ignore other approaches to maintaining environmental quality and sustainability. These alternative approaches are based on an understanding and adaptation of the mechanisms that ensure productivity and sustainability in naturally occurring ecosystems. *Working with Nature* advocates cooperation with nature by understanding its functions and transforming them into services beneficial to humans. This book seeks to convey an understanding of how nature works, and how this understanding can lead to a less intrusive and, thus, more sustainable management of farms, forests, and grasslands.

An ecological approach to resource management meets with several criticisms, however. One such criticism is that it is labor-intensive and, therefore, not economically feasible in developed countries, except perhaps where premium prices are paid for "organically grown" produce. In many parts of the world, however, labor is cheap and chemical subsidies are expensive. In such "lesser developed" regions, ecological agriculture could be, and often is, a practical alternative to technological agriculture. Further, it is often in such regions that the need for less damaging resource management is most urgent.

Another criticism is that there is not sufficient arable land remaining to feed the world's growing population through ecological agriculture. This is not necessarily true. Most mainstream agricultural crops and timber are produced in monocultures. In mixed species plantations, polycultures, and agroforestry where production of trees and crops are mixed, less land is required to produce a given level of food and fiber than where species are planted separately in monocultures. When livestock is added to the mix, efficiency of land use can be even greater.

An even more general criticism is that ecological agriculture is in its infancy and, therefore, has a long way to go to prove itself as a feasible alternative to more mainstream agricultural practices.

Unfortunately, ecological agriculture is in its infancy largely because it is unfunded, except through a few small private foundations. In contrast, energy-intensive agriculture is heavily subsidized through research and extension in agricultural universities, and has a broad supportive base in agro-industrial and agro-technical businesses. Given the low level of support for ecological agriculture, it is not surprising that it is not very profitable.

In a review of a recent text on agroecology, Kass (1997) criticized the material for not answering the following questions: "Does agroecology improve the standard of living of its practitioners? How many countries have enough labor and land for a more extensive agriculture? Does agroecology bring about an improvement in human health?"

These are not fair questions. Before they can be answered, the ecological approach to agriculture, forestry, and range management must be given a chance to work out its problems, develop, and mature. It must be given the same support that "conventional agriculture" has been given for the past 140 years. Only then can ecological agriculture and forestry be fairly evaluated.

Working with Nature does not advocate abandoning current mainstream agriculture and forestry and substituting ecologically based resource management. Rather, it suggests a gradual replacement of energy-intensive resource management with management based on an understanding of nature, as that understanding begins to emerge. The book's basic proposition is that *taking advantage* of the services of nature has the potential to be more economically efficient, require less land, and be more profitable than *fighting against* the forces of nature — as is typically the case with energy-intensive management programs.

* * * *

I am grateful to Dr. Florencia Montagnini for her helpful suggestions on the manuscript; and I thank Dr. B. T. Kang for helping me adapt material in our agroforestry course for this book.

INTRODUCTION

Everyone agrees that "sustainable" management of our farms, forests, and rangelands is desirable; but it is hard to find a definition of sustainability on which everyone can agree. An extreme view is that the only kind of resource harvesting that is sustainable is that of aboriginal hunter-gatherers in the rain forest, who live in a closed ecosystem where nothing is artificially imported into the ecosystem, and where, therefore, there is no pollution. Nothing is transported out of the ecosystem, so there is no depletion. Everything is recycled, and natural soil erosion is so slow that losses are replaced by natural soil-forming processes. Such systems, however, can only support a limited human population, and only in areas with fertile soils and abundant rain.

In societies where most of the production is for export rather than for subsistence, this concept of sustainability is meaningless. Economically interdependent societies exist in open ecosystems and continually import material into their environment that ends up as pollution. Such societies continually export material with resulting depletion of resources. To have meaning for a global economic society, sustainability must be defined in modern economic terms. But modern economic time horizons are on the order of a few decades — an interval that is ridiculously short in light of the time required to restore forests and rangelands (hundreds of years) or soils (thousands of years).

It is easier to agree on what sustainability is *not*. Sustainability is *not*: eroding soils; streams contaminated by fertilizers; human health threatened by pesticides; air polluted with wastes; crops and forests ravaged by disease; rangelands overgrazed; exotic pests choking out native flora and fauna; biodiversity threatened by clear-cutting; and management techniques dependent upon exhaustible sources of ground water and fossil fuel energy.

Are erosion, pollution, overgrazing, pests, and declining biodiversity a problem for modern agriculture, forestry, and range management? Most mainstream agronomists, foresters, and range managers will admit that these problems do exist in some places. Some believe that these are not long-term problems because technical solutions are said to be at hand or just over the horizon (Avery 1995).

Is the dependence of resource management on exhaustible ground water and fossil fuel energy really a problem? Most mainstream economists have to admit that at the present rate of use, these resources will run out someday. Nonetheless, they have faith that as these resources become scarce, technical solutions will provide substitutes for these resources.

Technological optimists have a strong argument in their favor. They point out that since 1798 when the English political economist Thomas Malthus predicted that population when unchecked will outrace the earth's ability to provide sufficient food, Malthus and his followers have continually been wrong. The reason is that technological innovation has drastically increased the ability of humans to extract resources from the earth. Now, in a similar vein, when environmentalists predict that the increasing world economy will result in intolerable environmental consequences, the technologists counter that technology will come to the rescue.

It is true that technology has spawned solutions to resource management problems; however, each advance in technology has also had unforeseen or unwanted side effects. Because of small scale, the adverse consequences of early technological advances had a limited impact on society as a whole. For example, the moldboard plow improved immensely the farmer's capability to cultivate the soil. The downside of the moldboard plow — erosion and soil compaction — was limited to the farmer's field.

In recent decades, the downside impact of new technology has had more far reaching and rapidly appearing consequences. The negative impacts of higher intensity production are no longer limited to the farmer and his fields, but rather are spread over vast regions, thus affecting entire communities. Some of the reasons for these far reaching, rapidly appearing consequences are as follows: the scale of application has been much greater; the potency of the technology has increased; the time interval required for development of new technology has lessened.

Resource managers are caught in a tightening spiral of using more and more technology that subsequently causes secondary problems that require further technological "fixes." Many mainstream managers and economists see nothing wrong with this. Although problems emerge more rapidly, technology also continues to advance more rapidly.

But there is another weakness with dependence upon technology: human and technological fallibility. When management scale was small, human lapses and failures of technology were merely of local significance. A break in a dam that controlled a water wheel a century ago might have affected a few farms along a river. Today, however, failures can have global significance. Chernobyl, Bhopal and the Exxon Valdez are only a few examples of instances where lapses in control had far reaching and tragic consequences.

So we have a controversy. On the one side are technological optimists, who have faith that human innovation can keep the world comfortably ahead of the problems created by the previous round of technology. On the other side are those who fear that the increasing scale and increasing potency of technology, combined with the frailty of human judgment, pose a significant threat to sustainability of our farms, forests, and rangelands.

Working with Nature begins with evidence of the dangers inherent in our increasing reliance on intensively applied technology to solve the world's resource problems. The evidence is not conclusive. It does not prove that environmental disaster is imminent if we do not abandon our current reliance upon plastics and pesticides. The book suggests, however, that we adopt the "precautionary principle" in adopting new technologies. This principle advises that if we are not able to predict very well what the outcome of our actions will be, we should consider taking other, less risky alternatives.

Working with Nature presents a less risky alternative to the current increasing reliance upon exhaustible petroleum and fallible technology. It advocates an approach that is more reliant upon nature's services sustained by energy from the sun. Such an approach has the potential to be more sustainable and to be less susceptible to human failures and negligence. At the present time, such an alternative may lower the profits compared to current, exploitative techniques. If the same subsidies and guaranties currently given to energy intensive farming and forestry were to be given to a more sustainable approach, however, ecological resource management could become economically profitable.

CHAPTER 1

THE RATCHET EFFECT

Technology has increased the productive capacity of our farms, forests, and rangelands. It also has increased the standard of living in those regions able to afford modern technology. But technology has drawbacks as well as advantages, and the disadvantages often do not become apparent until the technology has become deeply integrated into our economy. By then, the technology has allowed the population to expand and increase its standard of living. These are trends that are difficult to reverse, and so we have a *ratchet* effect.

The ratchet analogy refers to a spring that is wound tighter and tighter, prevented from releasing its energy by a ratchet. Just as the ratchet only allows the spring to tighten, our increasing dependence upon technology forecloses our options to survive without technology; and like the spring, whose potential energy increases with each notch of the ratchet, the potential for calamity increases with growing dependence upon fallible technology.

THE MECHANICAL ROUND

In the United States, the first round of the ratchet effect began in the early 19th century. Prior to that, most American farmers were limited by labor — their own, or that of slaves or hired hands. Draft animals were their only subsidy. The ratchet began to turn with the emergence of the Industrial Revolution and the market economy in the 1800s. Such mechanical implements as plows, tractors, cultivators, and the cotton gin revolutionized agriculture. Steam-powered machinery enabled loggers to extract and mill trees at a rate much faster than was possible with animal traction, and in the 1900s, the chainsaw and diesel-powered bulldozer enabled loggers to increase rates of extraction by an order of magnitude.

Taking the Plains

Before the beginning of the nineteenth century, the herds of bison that roamed the great plains were a sustainable resource for native American peoples. Beginning in 1832, there began a series of technological innovations that, in conjunction with the introduction of cattle to the western ranges, changed the Plains ecosystems from a state of sustainability to one of dependence upon subsidy. One crucial technological innovation was well drilling, which replaced digging, and allowed tapping of the deep water tables. Another was the self-regulating wind pump, commonly called the windmill. It was capable of reliably drawing water up from the deep wells and supplied a non-supervised means of regulating flow. A third innovation was barbed wire, invented in 1873. It was the first means of fencing that effectively controlled cattle movement and could be easily erected in extensive grasslands.

It was this combination of well drilling, windmills, and barbed wire that made it possible for the land to be fenced into smaller areas and for the stockmen to cut their ranges into pastures. Thus began the transition from sustainability dependent upon regulation by natural cycles to unsustainablity dependent upon technology (McNaughton 1993).

Subsidies

Accompanying the introduction of mechanized resource extraction was a change in the rural economy. Mechanized agriculture allowed the farmer to become more efficient, when efficiency is defined as net economic profit per unit time of farmer effort. Mechanization, however, was economically feasible for the individual farmer only through specialization, so the predominant strategy was to replace a variety of subsistence crops with such commodity crops as corn and wheat. The timber and ranching industries also expanded in response to the growing market demand for their resources. As agriculture, logging, and ranching moved westward, the market economy followed them.

Meanwhile, the flood of immigrants to cities in eastern United States created a sharply increasing demand for food and fiber. To help farmers, ranchers, and loggers supply the needs of a growing country, state and federal governments established assistance programs. In 1862, the Morrill Act was passed to establish land grant agricultural colleges. Their primary missions were: to educate young people who were interested in agriculture; to research ways of increasing crop production; and to disseminate new information to farmers of the country through extension services (Smith 1971).

A few decades later, it became apparent that institutional mechanisms were also needed to ensure a continuing supply of forest products. To provide a source of fiber for the wood using industries, the federal government began the creation of forest reserves which later became National Forests. In 1900 there was the establishment of the Society of American Foresters, the beginning of professional forestry education, and the creation of the U.S. Forest Service (Gordon 1996).

Other resource extraction and development activities were also subsidized by the Federal Government. Ranchers were given low-cost grazing rights on public lands. The system of railroads that integrated farms and ranches into the national market also was encouraged by grants of free land to the railroad companies, and the killing of native American Indians by the U.S. Army was the essential subsidy that allowed development to proceed in the first place.

For the most part, the programs were successful. The growing population of America was supplied with food and fiber at a price so low that the economy developed far beyond provision for mere subsistence. Growth was important for farms, ranches, and logging operations — in the market economy, economies of scale gave the biggest advantage to the largest operators. The Department of Agriculture and the Forest Service saw their primary mission as helping to increase production. Since big farms could better utilize "modern" methods of increasing production, much of the government effort went toward assisting big farms in getting bigger, as well as increasing the scale of logging and ranching operations. As the economic importance of farmers grew, so did their political influence. The "farm block" became an important political influence in the mid-twentieth century. These states elected representatives who supported existing subsidies and fought for new ones.

Limits to Extraction

Subsidies for agriculture, forestry, and ranching — when first introduced — were a good thing. They enabled both the producers in the country and the consumers in the city to achieve a standard of living far higher than their ancestors had ever thought possible. The American Dream was based upon the exploitation of North America's resources — its rich soils and productive ecosystems. But the individual entrepreneurs did not develop these resources on their own. It was government subsidies that made development possible and brought great riches to those who most aggressively took advantage of the subsidies. Subsidies to increase production of food and fiber contributed to the initiation of the ratchet effect by increasing production — which, in turn, allowed the population to grow and the standard of living to increase.

Another factor contributing to the ratchet effect has been competition. As more and more producers entered the field and supply began to satisfy demand, producers had to minimize costs in order to win the competition for a share in the market. To do that, farmers and ranchers mined the soil and loggers mined the forests — that is, they extracted resources in the cheapest way possible, with little or no thought for replacement or sustainability. In the Midwestern prairies, this involved stripping the ground of its cover to plant annual grains; in the forest, clear cutting and burning without replanting; and on the range, overgrazing until the grass disappeared.

Few saw anything wrong with this system. Low prices were one of the things that made America great, and the competition that kept them low was applauded. There were, however, some who complained. At the beginning of the twentieth century, groups such as the Sierra Club organized and began to protest against the destruction of America's forests, but they had little effect on a national level. It took the dust bowl of the 1930s to jolt Americans into the realization that America's resources were not endless. As a result, the Federal Government established the Soil Conservation Service in 1935, and new methods of contour cultivation, windbreaks, and ground cover were encouraged and implemented, even though they often decreased short-term profit. Americans finally realized that sustainability had to be subsidized to be achieved just as high yield had to be subsidized to feed a growing America.

THE CHEMICAL ROUND

The next turn that tightened the ratchet began after World War II, with a vastly expanded economy that demanded greater production of food and fiber. The solution seemed to lie in chemical subsidies and crops that could respond to these subsidies. Chemical fertilizers greatly increased the potential for production. To achieve that increase, crops that responded vigorously to fertilizers were bred. But while the new lines of crops were highly productive, they were also more susceptible to disease and competition from weeds. The response was another kind of technology: pesticides and herbicides.

Like the increases in production brought about by the Industrial Revolution, increases in production brought about by the chemical revolution had a negative impact that was not felt until the beneficiaries of chemical agriculture, forestry, and range management had developed a strong economic constituency. This constituency formed a powerful political influence that to this day has often tried to discount evidence that chemicals in the environment can be harmful as well as helpful.

Pesticides

Humans can be exposed to pesticides in two ways: directly as a result of spraying; or indirectly, through eating food contaminated with pesticides. Some pesticides are similar in chemical structure to estrogens (Colborn *et al.* 1996), and when they are ingested, they can upset hormonal balance. Some researchers have suggested that they can contribute to such ills as breast cancer, testicular cancer, and a decrease in human sperm count (Kaiser 1996). Pesticides weaken the immune system to varying degrees, and there is evidence that, when exposed, even healthy humans are at an increased risk of infectious diseases and cancers (Repetto and Baliga 1996). The threat is especially high in developing countries where resistance is lower and exposure is higher (Repetto and Baliga 1996).

Poison in the Ecosystem and Laboratory

The poisoning of wildlife and reduction of bird populations also have been attributed to persistent pesticides. Fat-soluble pesticides such as DDT are concentrated as they pass up the food chain. Concentrations of DDT in individuals from the top trophic level in affected ecosystems were up to 10^6 times higher than those in the environment throughout the United States in the 1960s (Harrison *et al.* 1970). Species such as kestrels, falcons and eagles underwent reproductive failure. As it is passed up the food chain, DDT also is readily dispersed throughout the environment. High concentrations of DDT were found in both freshwater and marine fish, and detectable amounts were found in penguins and seals in the Antarctic and in the Bering Sea, far from any area where it had been applied (Buckley 1986).

Because of these apparent effects, persistent organochlorine insecticides such as DDT have been banned in the United States since 1972, and later in Western Europe and Japan. They have been replaced with others that degrade more quickly. While some scientists argue that even these new pesticides are dangerous, Avery (1995) claims there is no evidence that pesticides are harmful to humans or wildlife. The problem is that levels of pesticides found in the environment are usually lower than those needed to produce a reaction in experimental laboratory animals.

In the laboratory, pesticides have traditionally been tested one at a time, whereas in nature, humans and wildlife are exposed to many at once. There is a possibility of synergistic interactions, resulting in a danger to the immune system that cannot be predicted based on laboratory tests of single pesticides. Arnold *et al.* (1996) initially reported that when the pesticides dieldrin, endosulfan, toxaphene, and chlordane were tested in combination, activity shot up by a factor of 160 to 1600. However, they were unable to duplicate their results (McLachlan 1997). Complicating the problem of testing combinations of pesticides are the manifold uncertainties associated with risk assessment of pesticides, the length of time required to test the effects, and the economic pressure for the Environmental Protection Agency to approve a pesticide before long-term risks can be assessed (Wargo 1996). Even if effects of multiple pesticides are only additive, their increasing use will eventually threaten human health.

Farms

Equally controversial is the effect of pesticides on non-target insects. In the Rio Grande Valley of Texas, where cotton is an important crop, aerial spraying of malathion was initiated to vanquish the boll weevil. Scientists, however, have implicated the insecticide in the elimination of wasps and other beneficial insects that previously controlled other cotton pests, such as the armyworm. In 1995, many farmers suffered crop losses of hundreds of thousands of dollars, and in 1996, they voted by a margin of 3 to 1 to stop the program. Nevertheless, the Texas Department of Agriculture continued to assess farmers $12 to $18 per acre for spraying, whether they wanted it or not (The New York Times 1996).

Methyl bromide is a pesticide gas that is injected into the soil targeting insects, nematodes, and other pests that damage fruit and vegetable crops (Rauber 1996). It is cheap, effective, and extremely toxic. The Environmental Protection classifies it as a "category 1 acute toxin," the most dangerous type. It is also 50 times more destructive to ozone than chloroflourocarbons. Under the Clean Air Act, U.S. production of methyl bromide is to cease in the year 2001, and The Montreal Protocol on Substances that Deplete the Ozone Layer calls for a world ban by 2010. However, large agricultural interests such as California strawberry growers and Florida tomato growers are lobbying hard to be exempted from the ban, which they fear will cause them to lose market share. While alternative methods such as steam treatment are available, they are not as easy to use or as economical as methyl bromide (Annis and Waterford 1996).

Pesticide Aftermath

Today in the United States, there are more than 600 pesticides in use, and annual usage amounts to 2.2 billion pounds per year, roughly 8.8 pounds per capita. In 1991, the United States exported 4.1 million pounds of pesticides that had been banned, canceled, or voluntarily suspended for use within its borders. These exports included 40 million pounds of compounds known to be endocrine disrupters (Colborn *et al.* 1996). How much of this is returned to the United States in and on imported fruits, grains, and vegetables is unknown.

There is evidence that one pesticide no longer used in the United States is depleting populations of a migratory bird that summers on the Canadian prairies and winters in South America. In January 1996, ornithologists working on the Argentine pampas found 4000 dead Swainson's hawks shortly after the birds were exposed to the pesticide monocrotophos (brand name Nuvacron) used to control grasshoppers that fed upon sunflower plantations. The pesticide is available to pampas farmers from an international chemical company. The scientists estimated that 20,000 birds, or 5 percent of the world's Swainson's hawk population, perished in a single season (Line 1996). Similar effects can be expected in fish and domestic animal populations where feeding may be restricted to areas where pesticides have been applied.

Some undesirable aftereffects of insecticides include the resurgence of pests after treatment. When predator insects that feed upon herbivorous insects are killed by the pesticide, resurgence of the herbivores can be especially severe. Other effects include the elimination of economically beneficial insects such as honeybees (National Research Council 1989), and aesthetically pleasing insects such as butterflies (Longley and Sotherton 1997).

Herbicides

In *no-till* cultivation, seeds are planted by a machine that injects them into soil which has not been plowed or cultivated. This type of planting eliminates the need for plowing and harrowing, practices which expose mineral soil and cause erosion. In no-till cultivation, weeds are controlled by herbicides. Herbicides also have often been used to replace mechanical methods of weed control in forest plantations and on rangelands.

Although herbicides have reduced erosion due to mechanical control of weeds, they are not without problems. Herbicides used in the 1970s, such as 2,4,5-T, were easily contaminated with dioxin during its manufacture. It has been implicated in asymmetrical development of the brains in herons whose parents had fed downstream from a dioxin-spewing pulp mill (Henshel 1997). It is often difficult to prove any direct effects on adult experimental animals at levels found in the environment. However, the effects may be mostly indirect. While it took an almost lethal dose to impair the reproductive system in adult rats, even small doses did long-term damage to the reproductive systems of males exposed to dioxin in the womb and through their mother's milk (Colborn *et al.* 1996).

Persistence and Long-Term Effects

While modern herbicides appear to readily degrade in the soil, the rate of degradation depends upon climate, and physical and chemical properties of the soil. When herbicides are incompletely degraded, residues can contaminate drinking water supplies and water used for irrigation. Herbicide residues have been detected in surface water (Thurman *et al.* 1992) and groundwater (Burkart and Kolpin 1993) throughout the North American Midwest and elsewhere. Herbicides in drinking water supplies cannot be effectively removed by conventional treatment, or even by more sophisticated carbon filtration systems. In Iowa, 27 of the 33 public water supplies from surface water sources tested had two or more herbicides in treated drinking water samples (National Research Council 1989).

There is little evidence, however, that herbicides have any long-term effect on recovery of vegetation in sprayed areas. The long-term effects of herbicides on the forests of Vietnam were the subject of a conference held in Jan. 1983 in Ho Chi Minh City. While forests that had been sprayed suffered erosion and invasion by noxious plants, these effects were probably caused by peasant farmers moving into sprayed areas to cultivate crops. It was much easier for them to initiate cultivation in such areas than in areas where the trees had to be cut by hand (Jordan 1985).

Limits of "No-Till"

Evolution of resistance in weeds may be the most dangerous aspect of long-term use of herbicides. As a result of long-term exposure, some weed species are already resistant. For example, resistance to sulfonylurea herbicides has evolved in prickly lettuce (*Lactuca serriola*), kochia (*Kochia scoparia*), and Russian thistle (*Salsola* sp.) in and around winter wheat and along roadsides in the northwestern U.S. (Keeler *et al.* 1996). Consequently, new and/or more powerful herbicides may be required.

There are many gradations between no-till and conventional-till agriculture. Cultivation that involves plows and rakes with short teeth and light circular blades is often called minimum-tillage or conservation tillage. Minimum-tillage equipment that does not turn over the soil can be used in combination with herbicides to reduce the quantity of herbicides required.

Hormones

The treatment of dairy cattle with bovine growth hormone (BGH) to increase milk production has been vigorously promoted in the United States (Krimsky and Wrubel 1996). Some scientists argue that this synthetic hormone will be a boon to milk production, while others claim that it will cause a higher incidence of bovine disease and expose consumers to new health risks from contaminants in milk. The presence of estrogen-like compounds in foods raises the question of whether these substances pose a hazard to human health or to the development of babies. Female rats exposed to estrogens gave birth to babies that showed reduced ability to reproduce when they matured (Colborn 1996).

Health issues aside, it is not clear where the advantage lies in one cow that produces 6 gallons of milk per day over two that produce 3 gallons each. After all, the extra calcium that goes into the extra milk has to come from somewhere. The cow that produces 6 gallons has to ingest twice as much grass as those that produce 3 gallons each. So if a farmer has a pasture of 10 acres, he could stock it with 6 untreated cows that produce 3 gallons per day of hormone-free milk, or 3 hormone-injected cows that produce 6 gallons per day of hormone-tainted milk.

THE GENETIC ROUND

An alternative to increasing the potency of pesticides and herbicides and increasing the intensity of their application is to develop, through genetic engineering, varieties of plants and animals that are resistant to pests and weeds. This approach is an outgrowth of conventional plant and animal breeding.

Human-directed evolution began with the domestication of wild races. The process was slow and uncertain until the emergence of genetics as a science. As an understanding of the mechanisms of heredity became understood, geneticists were able to speed the process of breeding for desirable characteristics. Much of the effort in plant breeding went toward crops that gave high yields in response to fertilizers and irrigation. In animal husbandry, the goal often was to increase the production of meat.

First Steps in Genetics

Fighting Pests and Disease

In the quest to improve yield and to assure uniform quality, scientists produced genetically uniform populations. An inadvertent consequence of uniformity is that populations are more susceptible to disease than genetically diverse wild populations. Genetic uniformity increases epidemics because disease organisms are faced with only one defense. Once disease organisms evolve to overcome this defense, they can attack the entire population. As new, genetically uniform races of corn and wheat became increasingly responsive to fertilizers, they also became more vulnerable to attack. The solution was to breed new pest-resistant varieties.

The effort has been only partially successful. Wheat stem rust and European corn borer were once serious crop pests, but for several generations now, varieties have been introduced that are resistant to pests — at least for a while (National Research Council 1989). Eventually, however, the pest catches up in the evolutionary race, and new varieties are needed. The emergence of a virulent and fungicide-resistant strain of the Irish potato famine fungus *Phytophthora infestans* in the early 1990s is one example. It has been locally devastating, sometimes causing total crop loss and severe economic hardship for potato growers (Fry and Goodwin 1997).

Looking to the Past

One difficulty for geneticists fighting the evolutionary arms race has been that some of the genetic material for characteristics only recently recognized as important has been lost during breeding that focused solely on yield. Characteristics such as competitive ability against weeds and resistance to pests, cold, and drought were neglected. In many cases, they were thought to be unnecessary, because cultivation systems were supposed to supply those functions for the plants. When it became evident that human systems were sometimes inadequate, geneticists looked to wild varieties where defenses still existed. A classic example is the search for a perennial variety of corn. The advantage of perennials is their ability to withstand cold and drought, and to regrow without the necessity of yearly replanting that requires breaking the soil and intensive weed control. A search in the mountains of southwestern Mexico turned up teosinte, a primitive ancestor of modern corn that possesses tuber-like rhizomes that resprout at the beginning of each growing season (Iltis *et al.* 1979).

This primitive wild corn was a rare find. In many cases, original strains have become extinct, or the genetic material for the characteristics newly recognized as important may no longer exist — and even if it is available, the process of incorporating it into modern strains can be slow. The length of time required to breed a new variety has been a particularly intractable problem where the life-cycle is long, as in the case of trees.

Give-and-Take Genetics

Another reason that plant breeders have problems increasing both yield and resistance lies in the first law of thermodynamics: matter and energy cannot be created, only transformed. Therefore, any increase in yield must be compensated by a decrease in some other characteristic, such as disease resistance. The relationship of thermodynamics to genetics is based on the fact that the efficiency of photosynthesis is fixed — that is, the amount of solar energy captured by plants and converted into carbon compounds used by the plant has an upper limit. On a given area of land, there is only so much solar energy that can be captured by plants, and as a consequence, only so much carbon can be reduced. This carbon can be used in various ways: vegetative growth, production of fruit and seed, root growth to compete for nutrients and water, production of secondary plant chemicals that repel insects, and storage in roots and rhizomes. Under optimum conditions of water and nutrient availability, plants are limited in what they can do by the amount of sunlight that they can capture.

Geneticists do not increase the amount of solar energy fixed by photosynthesis. What they do instead is rearrange how it is used. When they select crops for high production of fruit or leaves and stems, they must select against nutrient acquisition, disease resistance, or other use of the photosynthate. For example, the energetic "cost" of increasing resistance in morning glory plants has been shown to equal a decrease in net production of leaves (Fineblum and Rausher 1995).

While scientists have not been able to increase the efficiency of photosynthesis, they have discovered that some plants have a naturally occurring photosynthetic cycle that is more efficient than the cycle found in the majority of plant species. This more efficient pathway is the so-called "C4" pathway and exists in certain warm climate grasses such as corn, sorghum, and sugar cane. It is not yet known if the improved efficiency of the C4 cycle can be bred into other important crop plants such as rice and wheat.

Genetic Engineering

Genetic engineering is a new approach to incorporating characteristics currently deemed desirable into living plants and animals. It has the potential to produce organisms useful for agriculture, forestry, food processing, pharmaceuticals, toxic waste degradation, and other economic purposes. One advantage of genetic engineering is that the genes for particular traits can be selected. In conventional sexual crosses, there is no control over the outcome — we take what we can get. In many cases, we know what we can get. If we cross a horse with a donkey, we always get a mule. While we have not yet produced a mule without the characteristic for stubbornness, genetic engineering holds that possibility.

To produce a genetically engineered species, it is not necessary to find a variety possessing the desired characteristics. Genetic material can be transplanted between unrelated species. In contrast to hybrids, genetically engineered organisms are produced by artificially replacing a particular gene — sex is not involved. If a genetic engineer wanted to create a yellow tomato, he or she could insert the "yellow color" gene from a daffodil, or even a butterfly (Rissler and Mellon 1996). Genetic engineers have already used recombinant technology to insert genes from plants, animals, bacteria, and viruses into crop plants. For example, genes from chickens and silk moths have been spliced into potatoes to confer resistance to bacterial diseases; genes from bacteria have been inserted into alfalfa to produce an oral vaccine against cholera; and genes from flounder have been introduced into tomatoes to reduce freezing damage (Union of Concerned Scientists 1993). The Monsanto Corporation has marketed a potato protected against the Colorado potato beetle by a gene derived from *Bacillus thuringiensis*, a bacterium which produces a natural insecticide (Wadman 1996).

Yet another advantage of genetic engineering is the speed at which new characteristics can be incorporated. In the case of forest trees, it is not necessary to wait until the tree matures and produces flowers to initiate genetic changes.

Risks of Genetic Engineering

In contrast to traditional breeding where combinations are limited to traits of plants that can sexually interbreed, transgenic breeding has an enormous pool of genes to draw upon. A completely foreign gene may interact with the rest of the plant's genetic structure in wholly unpredictable ways (Ninio 1983, Regal 1988). Traits that appear advantageous in the short term may prove undesirable from a long-term perspective. Rissler and Mellon (1996) warn that since combinations with harmful potential can be produced rapidly in the laboratory, there is the possibility of releasing a whole host of new species that can impose unforeseen disasters before long-term consequences are assessed. Plants engineered to produce potentially toxic substances could affect unintended victims. Plants that are toxic to insects are also likely to produce products that are toxic to humans, and drug-producing plants could poison birds feeding upon those plants. The genes that enable plants to resist disease also confer decay-resistance, resulting in accumulation of toxic residue that could poison beneficial soil organisms such as earthworms (Clark 1994).

Just as pest species have evolved resistance against pesticides, they can also evolve the ability to overcome pest resistance genetically engineered into plants (McGaughey and Wilson 1992, Tabashnik 1994). Computer simulations by Alstad and Andow (1995) suggest that patchworks of treated and untreated fields can delay the evolution of pesticide resistance. The theory is that insects without resistance will survive in large numbers in untreated fields, thereby delaying or preventing the evolution of pesticide resistance. This, of course, presumes that farmers will agree to maintain unresistant patches, even though those patches will sustain heavy pest damage.

Another idea to prevent the evolution of insects invulnerable to "pest resistant crops" is to require the dose of toxin in the plants to be high enough to kill almost all insects exposed to it, so that few survive to breed. Theoretically, any surviving insects would mate with susceptible insects from nearby areas of non-resistant crops, thereby diluting any resistant genes. This was the approach required by the U.S. Environmental Protection Agency for the introduction of two million acres of *Bt cotton* in Texas. The crop is named after *Bacillus thuringiensis*, genetic characteristics of which are introduced into the cotton to produce a natural insecticide. However, the effort was not very successful. In the summer of 1996, thousands of acres of Bt cotton in Texas was found to be infested with the bollworm (Macilwain

1996). Insect adaptation to Bt toxins occurs through multiple physiological mechanisms (Oppert *et al.* 1997), and is controlled by both recessive and dominant genes (Tabashnik *et al.* 1997), resulting in rapid adaptation to toxins.

Yet another approach involves the use of mixtures of Bt toxins within each plant. Resistance was delayed when a pest population was exposed to more than one toxin at a time (Paoletti and Pimentel 1996).

A further step in creating insect-resistant plants has been arming plants with a genetically engineered virus deadly to insect pests. Warnings have come from experts in RNA viruses that their tendency to mutate might allow them to jump to a new host, thereby eliminating the herbivory that keeps many weeds under control (Service 1996). What are the implications if all plants become insect-resistant and decay-resistant? It would mean the demise of all insects and decomposer organisms. Plants pollinated by insects would not reproduce. Beneficial bacteria such as those that fix nitrogen might be eliminated. Undecomposed litter would accumulate and lock up nutrients so that plants could not grow.

Herbicide resistance, as well as pesticide resistance, can spread in undesirable ways. Herbicide resistance genetically engineered into crop plants can spread into wild species of closely related weeds. Mikkelsen *et al.* (1996) studied the transfer of genes from oilseed rape into a weedy relative of the same genus. They found that transgenic, herbicide-tolerant weeds are produced as early as the first backcross generation. Such new species often lack or escape the natural controls which regulate the populations of naturally occurring species. Transgenic crops themselves can become weeds.

THE TIGHTENING RATCHET

The first round of agricultural technology — that is, the time between its implementation and the manifestation of serious side effects — spanned a millennium. Approximately 3000 years before the Christian era, wheat was an important crop in the city-state of Sumer, between the Tigris and Euphrates rivers. Although the climate was hot and dry, irrigation permitted a large surplus production. However, the upward movement of the water from depth between periods of irrigation carried dissolved salts to the surface where they accumulated and damaged the crops. Between 3500 and 2500 B.C.E., wheat production fell drastically. Consequently, the size of the bureaucracy and army that could be fed and maintained fell, making the state vulnerable to external conquest. The decline and fall of Sumer closely followed the decline of its agricultural base (Ponting 1990).

The round of fossil fuel–powered mechanical technology in the United States and Europe covered approximately a century, from the early steam-driven tractors to the huge bulldozers with root-rakes used to clear brush and trees. The chemical round lasted merely a generation, from after WWII to the 1980s. Now, genetic engineering is touted as the technology which will provide the basis for continued economic growth in agriculture, forestry, and animal husbandry. How long will it take before undesirable side effects of genetic engineering create serious environmental problems? We certainly cannot assume that genetic engineering will provide the ultimate solution to problems created by the economic pressure to increase rates of resource extraction. The only thing that we can say for sure is that a combination of an expanding economy and the competition to produce more for less will drive entrepreneurs to seek even more powerful controls over nature, thereby increasing civilization's dependence on technology.

CHAPTER 2

THE PROBLEM OF SCALE

INTERACTION WITH TECHNOLOGY

In spite of the public rhetoric about the desirability of sustainability, the management of resources is becoming less sustainable, not more. The ratchet effect is one force that drives research and management in the direction of unsustainablity. A second factor is the increasing *scale* of modern resource management. The lower short-term cost resulting from "efficiencies of scale" also drives the trend toward decreasing sustainability of management systems.

Problems of scale occur because a linear increase in the size of the area managed is accompanied by an exponential increase in the associated environmental problems. When increases in scale accompany intensified technology, the negative aspects of both are multiplied. In the case of genetic engineering for pest- and herbicide-resistant plants, scale plays an important factor in the rate at which the ratchet turns. The larger the area planted to pest-resistant plants, the more rapidly the pests will evolve to overcome that resistance. Similarly, the larger the area planted to herbicide-resistant crops, the higher the probability that herbicide resistance will spread into weed populations.

Another example is the genetic engineering of crops to expand their range into marginal environments. Scientists are trying to introduce genes into wheat to enable it to grow in regions for which it was previously not adapted (Moffat 1992). The problem with introducing crops into agriculturally marginal environments is that these regions are far more fragile and prone to disaster than areas of good soils and better climates, where wheat is normally cultivated. It will take less of a mishap, natural or human-induced, to disrupt these marginal systems and throw the increased millions of people dependent on them into critical conditions.

MEAT AND DAIRY PRODUCTION

Environmental pollution due to the increasing size of meat and dairy cow production is an important example of the problems of scale. In the early days of the United States, beef cattle would graze the open range where their manure was recycled. On small farms, the farmer would shovel manure from the cow shed or the pig sty into his wagon and then scatter it onto fields to help soil fertility. Since quantities were small and neighbors were far away, there were few, if any, complaints about odor and water pollution. Today however, only a small proportion of beef, dairy, and pork products are produced on small family farms or fattened on the open range. Economies and efficiencies of scale for the production of meat and dairy products have forced many small producers out of business.

An important efficiency of scale in milk and meat production occurred when human labor — for shoveling manure out of the barns, or sties on small farms — was replaced on large farms by automatic flush systems that washed the manure into holding ponds or lagoons (Fulhage 1994). The slurry is held in these ponds and then periodically pumped out and applied to the fields. Unfortunately, these waste lagoons and liquid manure application systems create a whole new host of problems. A resource conservationist who deals with animal waste problems in the state of New York wrote, "The most common complaint I receive relates to odors associated with the spreading of liquid manure" (Gillette 1994).

The most serious consequence of liquid manure is leakage or overflow from the storage ponds into streams or into groundwater. More than 400,000 people in Milwaukee became sick from the protozoan *Cryptosporidium parvum* that may have come from city water contaminated by runoff from manured fields (MacKenzie *et al.* 1994). *Giardia*, another protozoa which causes diarrhea, cramps, nausea, fever, vomiting and fatigue in humans, is also the culprit in many of the outbreaks of gastroenteritis in the United States (Pell *et al.* 1994).

Leakage of slurry from holding ponds or leaching from fields also affects streams. Ammonia nitrogen, a product of manure decomposition, occurs in storage ponds. Even in small amounts it can be toxic to fish. A manure basin accident in New York State in 1989 affected about 25 miles of creek, killing over 17,000 (Gillette 1994). Phosphorus is also a common problem on sandy soils and can contribute to eutrophication, a process in which lakes and ponds are choked with the luxuriant growth of water weeds. Leaching of nitrates from decomposing manure in beef-cattle feedlots can also cause problems if they contaminate drinking water. Elevated levels of nitrate pose a health hazard to pregnant women and babies, as consumption results in a chemical imbalance in the blood (Gillette 1994).

A common cause of leakage from storage lagoons is inadequate sealing. However, the cause is frequently human. Manure disposal is a disagreeable, time-consuming chore. The waste storage pond is emptied only when it is close to running over. Rates of application to fields are often governed more by the quantity to be disposed of rather than by the capacity of the soil to absorb the byproducts of decomposition (Gillette 1994). The problem is not that the technology to solve the problems of manure pollution does not exist — it is that it is inadequately applied.

The economies and efficiencies of scale that bankrupted the small meat producer and dairy farmer were based on an economy that only took into account the costs of feeding and maintaining the animals and the marketing of their products. The economic reckoning did not take the environmental costs into account, as they were assumed to be zero. Said New York State Commissioner of Agriculture R. T. McGuire (1994) in his plenary address to a 1994 symposium on Liquid Manure Application Systems, "In the past, we assumed too often that when manure went onto a field it stayed there. If there was runoff, we considered it harmless."

Because of the order-of-magnitude increase in the scale of meat and dairy operations and the higher concentration of animals in feed lots and dairy barns, environmental costs are no longer negligible. The immediate solution appears to be the establishment and enforcement of more rigid regulations, and more scrutiny of operations by government inspectors. Such measures have a monetary cost in the form of salaries for officials who make and enforce the regulations.

However, these costs are borne almost entirely by the general public, not by the offending industry. So it is the community as a whole that suffers diseconomies of scale, not the individual producers.

SCALE AND FORESTRY

While pesticides and herbicides are often used in forestry operations, it has been the large scale of operations in combination with heavy-duty technology that has been causing environmental problems not only in the United States, but in much of the world. Early loggers cut down one tree at a time and used animal traction to haul logs from the forest. Then, by the turn of the century, there was a transition to steam tractors, and later, bulldozers with skidders to haul logs from the woods. Because of the expense of the heavy equipment, clear-cutting was more economical than selective cutting. It was cheaper to cut down or bulldoze all the trees in an area than it was to try to manipulate the huge machines across rugged terrain and between trees. Further, with clear-cutting, replanting for an even-aged monoculture was greatly simplified. By the 1960s, complete harvesting of existing stands — often by clear-cutting, followed by reforestation with seedlings from nurseries — became standard practice (Gordon 1996). When environmental impacts were not included in the calculation of costs, the system often appeared profitable.

The problems with clear-cutting have been erosion that chokes adjacent streams (Tamm *et al.* 1974) and the destruction of habitat for rare and endangered species (Salwasser 1986). Another problem is that the even-aged monocultures that often follow clear-cutting are highly susceptible to insects, such as the pine bark beetle and the spruce budworm. The spread of disease through plantation monocultures can be particularly severe in the tropics, due to year-round conditions favorable for infection (Jordan 1985).

In recent years, the increasing scale of clear-cut operations in national forests has caused conflict with the increasing demand for recreational use of national forests. When the national forests were established in the United States, the intention was that they should be used for both a supply of timber and a source of recreation. Before the 1960s, when clear-cutting became common as a result of economic pressures and powerful technology, logging and recreation coexisted without major conflicts. As both have increased in scale, however, clear-cut logging and recreation are becoming increasingly incompatible.

IRRIGATION AND SCALE

One of the more striking examples of expansion of scale at the cost of long-term sustainability is the exploitation of the Ogallala aquifer, which underlies 174,000 square miles of fertile but dry plains stretching from Texas to South Dakota. More than a half-billion acre-feet (the amount of water needed to cover an acre to a depth of one foot) of water was used between 1960 and 1990 to irrigate what was once the Dust Bowl. Pumping and irrigation technologies have made this area one of the largest and most productive farming regions of the world. Unfortunately, it is nonrenewable. It is essentially fossil water that percolated down from glaciers in the Rocky Mountains tens of thousands of years ago before it was geologically cut off by the Pecos River and the Rio Grande. By the early 1990s, more than one third of the Ogallala water was used up (Opie 1993).

Some hydrologists predict that at the current rate of drawdown, the aquifer will fail to yield in another 40 years (Ferguson 1983). The effects of the reservoir's exhaustion is already apparent. Land in Texas, New Mexico and Kansas is being taken out of production because of diminishing well yields and rising pumping costs (Ferguson 1983). Stokes (1983) predicted that 15 million acres in 11 states may be taken out of production by the year 2000. More efficient irrigation systems may slow the aquifer's drawdown, but ultimately, complete depletion is inevitable, unless the goal of "zero depletion" set by the Northwest Kansas Groundwater Management District is met. However, over-investment in irrigation equipment is making this goal extremely difficult to attain (Opie 1993).

SCALE AND LOSS OF SPECIES DIVERSITY

Availability of wild genetic stock can be important if genetically uniform plantations display some unforseen vulnerability, or if genetic engineers suddenly see a need for a particular gene found only in wild populations. Unfortunately, wild species are rapidly becoming extinct. The World Wildlife Fund (United States Chapter) estimates that at least 480 species of native plants and animals are known to have vanished in the United States in the last 200 years, including the ivory-billed woodpecker, the Florida black wolf, the passenger pigeon, the Arizona jaguar, and the Carolina parakeet. Currently, over 4000 species and subspecies are recognized as candidates for endangered species status within the United States. In many other nations, especially in the tropics, a greater number of extinctions have occurred. Biogeographers estimate that by the year 2100, 25 to 50 percent or more of tropical species will vanish (Soulé 1991).

Loss of habitat due to conversion for farming, forestry, and grazing is rapidly eroding the wild genetic stock that contains characteristics that could save our forests and our agriculture.

CHAPTER 3

AN ALTERNATIVE PHILOSOPHY

What can we do about the problems that result from the intensification of resource management and the expanding scale that often accompanies it? We can ignore them, but that only works for a while. A second approach is to develop new production technologies that are not as environmentally detrimental or that mitigate the effect of detrimental technologies. Many colleges of agriculture and forestry have recently broadened their teaching and research from a narrow focus on increased production to one that includes environmental problem mitigation. But most of the new technologies are susceptible to the same dangers as earlier technologies designed solely to increase production — they are subject to fallible human control, they may be expensive, and they may have unanticipated side effects. The economy is allowed to expand, but when the problems appear, the economic ratchet has foreclosed the possibility of turning back. The economy cannot contract or even remain in a steady state without major political repercussions.

The third approach incorporates elements of what has been called "alternative agriculture," "sustainable and holistic resource management," "ecological and organic agriculture," "permaculture," "low input agriculture," "agroecology," and "agroforestry." It espouses cooperating with nature, rather than conquering her, in order to sustain her productive capacity.

WORKING WITH NATURE

"Working with nature" is an approach to resource management that differs philosophically from that based upon the assumption that nature must be conquered, controlled, and disciplined to suit perceived human needs. That philosophy was successful on the frontier, when immediate survival of humanity depended upon conquering nature. However, the ability of humanity to survive for the short term is no longer in doubt — the question is whether humanity can survive for the long term. The answer depends in part on whether we will continue to rely upon technology-intensive management that has yet to prove that it can be as effective at sustainability as it is at short-term profit making.

Working with Nature takes the view that resource management should be based on an understanding of the naturally occurring interactions within ecosystems and how these interactions can be used to provide for human needs. It includes the study of natural interactions between different species of plants, between plants and animals, and how these interactions can be substituted for technological control. In practice, we first try to understand the interactions and then design a system that takes advantage of what nature is already doing. For example, we look at the ecological services performed by trees, such as recycling nutrients and controlling weeds. Then, we design an agroforestry system in which interplanted crop plants can take advantage of these services.

Another example of the contrast between conventional and alternative resource management is in the area of pest control. Until about a decade ago, a frequent approach was to spray a field with insecticides until no more insects could be found. The alternative approach recommends that pest management be based on reliance on natural functions, such as wasps that are parasitic on pest species. The challenge for ecological scientists interested in sustainable resource management is finding the right predator to control the pest, and determining what type of habitat is required by the desired predator.

ECONOMIC FEASIBILITY

Is it feasible to change the way we manage our resource systems — farms, forests, and rangelands — from conquest to cooperation?

Taking advantage of the services of nature is often not only labor intensive, but intensive in information and understanding of the functioning of natural systems. Standing in the way of its adoption is the fact that labor is expensive in the Unites States, and the ecological science upon which natural management is based is not readily available to the managers of agribusiness, forest products industries, and meat production operations.

Most colleges of Agriculture, Forestry, and Range Management have tried to jump onto the environmental bandwagon. Philosophically, however, they still adhere to an economically based mentality that bigger is better, and that high-tech is better than low-tech. Thus, bigger recycling systems are better than smaller recycling systems, and insect control based upon chemical baiting is better than insect control based upon diversification of crops.

In contrast, the idea of working with — instead of against — nature is an ecologically based philosophy, where the economic system accommodates the forces of nature rather than vice versa. Mainstream economists criticize such an idea as being unrealistic and impractical. What critics forget is that agriculture — as practiced in the United States 150 years ago — was also unrealistic and impractical for economic development and progress, and forestry and range management did not exist at all. To stimulate the development of resource using industries, the government had to subsidize farming, forestry, and ranching. Only now in the 1990s, as sustainability is competing with increasing production as a goal of resource management, are some of the subsidies for production finally being phased out.

Now that sustainability has become an important goal in resource management, it is time to begin work on systems that are truly sustainable, as long-term sustainability can be achieved only when reliance on the services of nature begins to be substituted for reliance on human technology. But where will the movement toward more sustainable agriculture begin? Not with industry — the world's agrobusinessmen are unlikely to suddenly become organic gardeners. Their responsibility is to make a profit for shareholders. Who, then, should *be* concerned with developing sustainable agriculture?

There are a few privately funded organizations, such as the Rodale Institute and Winrock International, that are concerned with long-term agricultural sustainability. In the public sector, it is only the universities that have the freedom and opportunity to pursue interests that are not justified by economic profit.

However, there is little support for alternative agriculture and agroforestry within present day research universities because of the little support given by government funding agencies. The conversion to sustainability can begin only when governments are convinced that serious subsidies must be dedicated to not only encourage research on alternative agriculture and forestry in the universities, but to encourage their practice in the private sector.

CONCLUSION

Nature is no longer a wild frontier, and research need no longer focus on conquest. Nature still can provide much to humans, but little is to be gained by further strong-arm tactics. The challenge for research universities at the beginning of the twenty-first century is to learn how to meet human needs through the cooperation with nature, instead of the conquest of nature. Nature cannot be destroyed any further without destroying the human species reliant upon it. Our new relationship with Nature should be one of sustainability. In order to begin a mutualistic relationship with nature, we have to understand how nature works, refrain from killing the goose that lays the golden eggs, and instead, discover how to nurture it before it dies from abuse.

CHAPTER 4

THE BOTTOMS-UP APPROACH

"Working with nature" is based upon an understanding of the many interactions that occur in nature, and using these interactions wherever possible to replace technological functions such as mechanical weeding, chemical pest control, and inorganic fertilization. There are two approaches to understanding the relationship between how nature functions, and how humans can take advantage of these functions for their resource systems — the *bottoms-up* approach and the *top-down* approach.

The two approaches can be contrasted by making an analogy to understanding an automobile. In the bottoms-up approach, one would begin by analyzing the system components such as the carburetor, fuel pump, radiator etc. and the interactions between components. Based on the characteristics of parts and their interactions, one predicts the performance of the entire system, the car. In the top-down approach, one begins by determining the performance of the entire system (acceleration, speed, load capacity). Through an analysis of the system performance, one predicts the nature of the components and their interactions underlying the system.

We begin our analysis of agro-ecosystems with the bottoms-up approach. In the context of an ecosystem, bottoms-up begins with an analysis of ecosystem components and the interactions that take place between them. Once some of these interactions are understood for a particular ecosystem, a model is conceptualized that synthesizes the interactions — or at least the relevant ones within the system. This model then is the basis for design of a resource production system (farm, forest, pasture) based on the services of nature.

Species or sometimes individual organisms can be the primary components in ecosystem models. Often it is useful to consider functional groups such as trophic levels as basic components. Much is already known about the characteristics of ecosystem components. Much less is known and understood about the *interactions* between components. The next several chapters deal with interactions between system components, because it is these interactions that determine the performance of the entire system.

Interactions between individuals, species, or functional groups in ecosystems have been classified in various ways. One of the oldest terms for a cooperative interaction between two species is *symbiosis*. Another term for interactions that benefit the interacting species is *mutualism*. Boucher (1985) has classified mutualisms into several categories:

1. **Nutrition/Digestion**. Increasing the availability of nutrients, energy and water to one or both of the interacting species. An example is *mycorrhizae*, an association between plants and a fungus through which the fungus obtains carbon for energy, and the plants ability to take up nutrients from the soil is increased.

2. **Protection**. Protection for one or both interacting species. Certain species of ants protect the plants in which they live. When an animal begins to browse a protected plant, the ants swarm out and sting the intruder.

3. **Pollination**. Pollination of some plants depends upon the actions of specialized insects and birds. The latter benefit from the mutualism by obtaining energy from the plant's nectar.

4. **Seed Dispersal**. The spreading of seeds away from the parent tree by birds, bats, and small mammals that eat fruits increases the probability that some seeds will survive and grow.

A positive interaction between crop species has been called *facilitation* (Vandermeer 1990). The term has been used when two crop species planted together (such as corn and beans) give a higher yield than when the same species are planted separately (Francis 1986).

To take advantage of the services of nature, the farmer or forester is not restricted to interactions that are mutually beneficial. There are other types of interactions that can be used to attain management ends (Lewis 1985). Some interactions are beneficial to one species but neutral to another (commensalism), or beneficial to one and disadvantageous to another (agonism). Negative interactions such as competition can be just as useful to the resource manager. The competitive ability of the crop itself can be used to suppress weeds through dense planting of the crop plants (Francis 1989).

Common interactions in ecosystems are those that occur as a result of food chain dynamics: the insect eats the plant; the bird eats the insect; the snake eats the bird; the snake eventually dies; the fungi decompose the remains; the bacteria break down the fungi; and when the bacteria die, their nutrients are recycled by the plant. The sustainable functioning of an ecosystem depends upon the integrity of its food chains.

When the bird eats the insect, a *direct* interaction takes place. However, indirect effects can be as important, or more important than direct effects, in governing the productivity and sustainability of an ecosystem (Patten 1991). For example, wolves can regulate the vegetative growth in an ecosystem through their control of the moose herd that grazes upon that vegetation. Both direct and indirect interactions within an ecosystem are important for maintaining that system's productivity and sustainability.

Interactions that ensure the integrity of the ecosystem often benefit all the participating *species*, although negative interactions may not benefit some of the *individuals* involved. The wolves benefit the moose herd by preventing overgrazing and eventual starvation, even though the individual moose that is eaten does not benefit. From the viewpoint of long-term sustainability of species, all interactions within a stable ecosystem can thus be considered mutualisms.

Our objective is to understand how interactions, or services of nature, can substitute for mechanical, chemical, or genetic controls in economic ecosystems — farms, forests, and rangelands. In the next four chapters, we will discuss interactions between plants, soils, herbivores, predators, and decomposers; how the interactions help increase productivity and sustainability in natural systems; and how they can be used instead of artificial subsidies in economic production systems.

CHAPTER 5

PLANT–PLANT INTERACTIONS

Individual plants within a community can interact in a variety of ways: some negative, some positive.

NEGATIVE INTERACTIONS

Competition

Most croplands in developed countries are monocultures, for economic reasons. Monocultures — or fields in which only a single commodity (such as corn, wheat, potatoes or cotton) is planted — are cheaper and easier to cultivate and harvest with mechanical and chemical means than are mixtures of species. The reason is that each species has unique requirements for production. With the specialized equipment necessary for large-scale production of each species and the different physiologies and phenologies of each species, it is difficult to raise several species at the same time in the same field.

In forest plantations, monocultures are also the norm. Because forest harvesting often requires the building of roads which is an expensive task, it is economically more profitable if the forest can be clear-cut. Planting is also an important expense in forestry, so most plantations are *even-aged* — that is, all the trees are planted at the same time.

There is an important disadvantage to even-aged monocultures — competition. Since all of the individuals are the same age and of the same species, they all have identical requirements at the same time. Their leaves are all at the same level and angle, so competition for sunlight at canopy level is intense. Their roots are all at the same depth as well, so competition for nutrients and water in the rooting zone is just as fierce.

Agronomists and foresters devote much of their attention to the problem of competition. The challenge is to plant individuals far enough apart to minimize competition for water, sunlight, and nutrients, while at the same time, planting them close enough together so that none of these resources are underutilized. When resources are underutilized, weeds will readily invade and exploit the unused resources.

Once they become established, weeds can often out-compete crop plants for light, nutrients, and water. Weeds are usually better competitors than are crops. In crop plants, the ability to compete in the wild has been sacrificed by plant breeders for other traits such as high productivity. Traits of wild plants that make them better competitors than domesticated plants include:

> larger roots; efficient association between roots and mycorrhizal fungi; production of root exudates that improve nutrient recycling; longer life span which allows nutrient storage during shortages; long-lived leaves which reduce the necessity of replacements; high nutrient use efficiency (ability to grow at low nutrient concentrations); strategic reproductive patterns that take advantage of natural seed dispersal processes and avoid seed predation; an ability to grow on acid soils; an ability to withstand drought; high resistance to insects and disease; and allelopathic repression of other plants. (Jordan 1985)

Determining optimal spacing so as to minimize competition between individuals of desirable species, while at the same time occupying as much land as possible to minimize opportunities for weeds to invade the cropland, is often difficult. The two goals require opposing strategies. Optimization of spacing is especially difficult in forestry, because of the long life of trees. If trees are planted far apart, e.g. five meters, then most of the sunlight and soil water and nutrients in a plantation are wasted for several years, until the tree canopy closes — that is, extends over the space between trees. If trees are planted close together, e.g. two meters, they are competing severely with one another by the time the trees are five or six years old and growth stagnates. The stand must be thinned, which is usually an economically wasteful operation. Another factor the forester takes into account in spacing is its effect on the shape of trees. Trees planted closely together shade each other from the side, thereby reducing growth of branches and stimulating vertical growth. Such growth results in trunks that are more valuable for lumber and veneer.

Genetic Solutions

In mainstream agriculture, forestry, and range management, weeds are controlled mechanically or chemically. In order to eliminate the side effects of these treatments, scientists are looking at other approaches to weed control. Genetic engineers are trying to solve the problem of weed competition by breeding crops that are better competitors (Lotz *et al.* 1995). However, in order to increase competitive ability, it is necessary to decrease yield. As pointed out in Chapter One, yield is limited by the amount of incoming solar energy that can be captured by photosynthesis. Genetically tinkering with plants has not been able to increase the efficiency of photosynthesis, but only redistributing the amount already available and changing the way it is used.

By decreasing the amount of solar energy that a plant uses in the struggle for water, nutrients, and light, the geneticist increased the amount available for yield. A good competitive ability was not necessary, because it was assumed that the farmer would supply the necessary nutrients and water, and eliminate the weeds that competed for space. Now, however, when the goal is to reduce the inputs to the farm in order to decrease environmental problems, the geneticist wants to breed back into the plant an increased competitive ability. But competitive ability requires energy, and since energy and matter cannot be created, only converted, the farmer must sacrifice yield in order to increase competitive ability.

Alternatives to Mechanical and Chemical Control

Organic farmers frequently spread mulch on top of the soil between crop plants to prevent establishment of weeds. Mulch suppresses weeds in several ways. It prevents light from striking the soil and stimulating the germination of weeds; it physically blocks the upward growth of germinated weed seeds and prevents them from getting sunlight; it prevents weed seeds blown into the field from reaching mineral soil necessary for successful germination.

In addition to weed control, mulch also conserves soil moisture by decreasing evaporation from the soil surface, and improves soil structure through addition of carbon. Mulch can also increase soil fertility when mixed with a source of high nitrogen, such as manure. However, mulching may be less beneficial at high latitudes, where it can prevent the soil from warming up and seeds from germinating. Mulching also may favor the spread of fungal diseases.

Mulch can be any readily available organic material or byproduct such as hay and straw. In the southern United States, organic farmers frequently use hulls of pecans and of peanuts as mulch. Wood chips are also an excellent mulch. An important disadvantage with these type of mulches is that they must be collected and spread — two operations which can be time-consuming and expensive.

In the Piedmont region of Georgia, we are experimenting with a rotation which may overcome the problem of collecting and redistributing mulch on grain crops. In the fall, crimson clover (*Trifolium incarnatum*) is planted, and a thick cover becomes established before severely cold weather occurs. When the warm weather begins toward the end of March, it is already growing vigorously. It flowers in late April, sets seed in early May, and dies back by the end of May, forming a dense, nitrogen-rich cover that effectively suppresses weeds and improves soil microclimate. Grain crops such as sorghum (*Sorghum bicolor*) are then planted through the mulch with a no-till planter. By the time the mulch cover decomposes, the grain is well ahead of the weeds. Little or no fertilizer is needed because of the nutrient supplying capability of the clover mulch (Matta Machado and Jordan 1995).

A common feature of pre-herbicide farming systems was the use of *stale seed-beds*, where the soil is cultivated and the resulting flush of weeds is killed off by further cultivation before the crop is planted (Cussans 1995). By the time new weeds emerge, the crop is already well established. Other techniques include soil tillage in darkness and burning off of weeds before emergence of crops (Rasmussen and Ascard 1995).

Another concept helpful in lessening the weed problem is that it is not necessary to eliminate all weeds. In some cases, weeds can have a beneficial effect (Altieri 1995). Weeds store nutrients and prevent their loss through leaching and erosion during winter or dry season, when no crops cover the soil. Decomposing weeds on the soil surface provide an energy source for beneficial soil organisms. If weeds are eliminated, pests that attack weeds may then attack crop plants.

Insects that use weeds as alternate food are reduced, thereby lessening natural control of weeds.

It is important to compare the cost of weed control with the decrease in yield due to weed control. There is an optimum point of weed control, beyond which the cost of further control outweighs the increase in profit resulting from additional control (Weiner 1990). However, this "economic threshold" is difficult to gauge, and an optimum threshold for a single season may be different than the threshold for long-term weed control (Hughes and Gonzalez 1997). An economically optimum threshold for a single season may result in a seed bank of weed species that is suboptimum over the course of several years.

Fallows for Weed Control

Fallowing means leaving land idle for one or more growing seasons. Long-term fallows that allow trees to mature result in weed control, because the closed canopy shades out many of the weeds detrimental to agriculture. In developing countries where shifting cultivation is still practiced, the elimination of weeds by trees is an important reason for fallow. After the trees are cut, the peasant farmer has two or three years to cultivate his crops before the thickening weeds make it more practical to clear new forest. Because of the large amount of land required by a system of long-term fallows, it is not practical in a modern economy.

Allelopathy

Allelopathy is a botanical term that is applied to biochemical interactions between plants that have detrimental effects on at least one of the interacting species (Chou 1990). Trees commonly used in plantations such as *Eucalyptus* spp. and *Leucaena leucocephala* often produce toxins that inhibit the growth and decrease yields of some (but not all) crops interplanted with, or subsequent to, these species (del Moral and Muller 1970, Suresh and Rai 1987). However, allelopathic effects can also be beneficial, as they are when used as a method of controlling weeds. Aqueous extracts of the forage crop Pangola grass (*Digitaria decumbens*) have been shown to be highly phytotoxic. With sufficient fertilizer, pangola grass forms a pure stand where almost no other weeds can grow (Chou 1990). Gliessman (1983) found that extracts of squash leaf inhibited various Central American plants, but not those of corn and beans with which it is traditionally interplanted. However, only a few studies have employed the allelopathic effect as a practical means of directly controlling weeds (Putnam and Duke 1974).

POSITIVE INTERACTIONS

Systems where positive interactions predominate are often more sustainable and/or more productive because less artificial subsidy is required, and environmental resources are used more efficiently.

Facilitation and Complementarity

Overyielding

When several crops interplanted together give a total yield per acre that is higher than when the crops are planted separately, the phenomena is called overyielding. For example, suppose that corn, beans, and squash — when planted in three separate monocultures — each yield 100 kg/ha for a total of 300 kg of yield over 3 ha. When planted together, however, each species yields 50 kg/ha, for a total yield of 150 kg/ha and 450 kg over 3 ha.

Overyielding may result from more efficient use of resources — land, nutrients, water, and sunlight — because the various species complement each other in the way they use these resources (Trenbath 1986). Some species have roots concentrated near the surface, while others have deep roots; when they are planted together, one can take up nutrients and water unavailable to the other. Different species often have different nutrient requirements; one species may need high levels of nitrogen, while another may be a calcium accumulator. Different species have different water requirements, and a combination of one water-demanding species with another needing less water can better survive dry periods than a monocrop of water-demanding species. Each species has its own characteristic leaf shapes and leaf angles. As a result, some are more efficient in the overstory, while some are more efficient in the understory. Some are more efficient using diffuse light, as occurs on a cloudy day, than in using direct sunlight. Each species has a distinct and separate niche. When overyielding is caused by a more efficient use of resources, we have *ecological complementarity.* We could also say that overyielding in mixed-species cropping systems results from a lessening of the intraspecies competition that occurs in a monoculture, where adjacent individuals all compete for the same resource at the same time.

Intercropping

Intercropping means using a field for more than one crop species. Intercropped species may not always exhibit complementarity, and when they do, the species may not complement each other in the use of all resources. In Costa Rica, several crop species planted together exploited the soil more completely than when each crop was planted separately, and were more effective in intercepting light than a maize monoculture. However, a sweet potato monoculture was as effective in intercepting solar radiation and reducing the impact of rainfall as more diverse ecosystems (Ewel *et al.* 1982.)

It is often difficult to distinguish between facilitation and complementarity as causes for overyielding. The corn–beans–squash intercrop provides such an example. Their use of light exhibits complementarity. The corn is more efficient in intercepting light entering the field from an angle closer to horizontal, whereas the squash — because its leaves lie flat on the soil surface — is more efficient with vertical rays. The beans climb the corn, and their horizontal leaves are dispersed along a vertical axis, similar to that of some trees.

However, facilitation also may be occurring in that the beans have a symbiotic relationship with nitrogen-fixing bacteria, and that nitrogen becomes available to the corn and squash. And because the corn stalks provide the beans with a pole upon which to grow, the corn facilitates the beans.

Another interaction that is facilitative is between pigeon pea (*Cajanus cajan*) and sorghum (*Sorghum bicolor*) in many tropical soils where most of the phosphorus is bound in insoluble iron and aluminum compounds. Guedes (1993) found that sorghum grown together with pigeon pea in an Ultisol grew significantly faster than sorghum grown alone, when the roots were in close proximity. Experiments by Ae *et al.* (1990) suggested that organic acids produced by pigeon pea solubilize phosphorus bound in the soil and render it available to the sorghum.

While overyielding is sometimes a benefit of facilitation and/or complementarity, sustainability often is the most important benefit. An example is the combination of overstory leguminous trees (such as *Inga* spp.) with understory crop species (such as coffee). The leguminous trees fix nitrogen, which becomes available to the coffee trees through leaf litter fall and root sloughing. The overstory trees may have deeper roots that can bring up and recycle nutrients (such as calcium and potassium) that have leached down to lower soil horizons. The shade trees also suppress weeds. While plantations of shade-grown coffee have a lower yield than sun-grown coffee, they require much less fertilization and weed control, and thus are less dependent upon external subsidies.

INTERCROPPING IN AGRICULTURE

To take advantage of positive interactions between crop species, the farmer must practice intercropping. The various species can be mixed spatially, planted sequentially in the same field, or both.

Double cropping is a kind of intercropping. Long-term crops, such as cassava, photoperiod-sensitive sorghum, and sugar cane, initially grow slowly and establish a full canopy only after several months. This provides an opportunity for interplanting of shorter-cycle crops such as maize, beans, cowpeas, and soybeans (Francis 1989). Intercropping cassava with maize in Nigeria resulted in a microenvironment which was cooler, higher in soil moisture, and with improved earthworm activity and water infiltration (Olasantan *et al.* 1996). The yield of maize was an additional benefit.

Interplanting agricultural crops between rows of *Paulownia* (a valuable, fast-growing timber tree) increased yields on summer harvested crops on the north China plain. Hot, dry winds are common there in the summer, and the trees protect the crops from the direct effects of the wind. The light shade cast by the canopy also appears to improve the moisture relations in the soil (Huo 1992).

An extreme example of intercropped species exists in the Kandyan Forest Gardens of Indonesia. There, Perera and Rajapakse (1991) found 61 species of trees, 12 of shrubs, 10 of climbers, and 42 herbaceous species in 50 of the gardens they sampled. The highest canopy layer was usually dominated by jack fruit and coconut. The next was usually arecanut and gliricidia which are used for wood. Coffee was the most common in the third canopy layer. Other important species interplanted in the gardens included cloves, avocado, mango, pepper, fishtail palm, and papaya.

Intercrop systems where alternate individuals are of different species are more frequently practiced in regions where labor is cheap, because machinery is generally not adaptable for such mixed-species cultivation. However, some of the advantages of intercropping can be obtained by row or strip intercropping, where rows or strips of one crop are alternated with rows of another. For example, when two rows of sorghum were alternated with a single row of pigeon peas, yield was significantly higher than when the crops were planted in monoculture. The reason was that more of the total growing season was used and more total radiation was intercepted (Francis 1989).

Another major advantage of multiple cropping is the reduction of risk. For example, in the event of a corn blight, a farmer who is cultivating maize along with cassava is better off than one who has planted maize alone. Even in the absence of disease, cultivating a variety of crops is a good strategy against variability in market prices. Yet another economic advantage is the steadier income stream derived from a succession of crops compared to the one payoff from a monoculture.

Alley Cropping

In tropical countries, soil organic matter usually decomposes quickly. Mulch is especially important to maintain soil quality (Jordan 1985). In order to obtain a more sustainable source of mulch, some tropical farmers use what is called "alley cropping" (Kang *et al.* 1995). In alley cropping, fast-growing leguminous shrubs are planted in "hedges." The grain crop, or other commercial crop is planted in "alleys" between the hedges. The alleys usually are about four meters wide. When the branches of the shrub grow and extend over the alleys, they are lopped off and allowed to fall onto the soil. The leaf mulch improves the microclimate, suppresses weeds, and increases nitrogen for the grain crops. In some regions, the leaves of the leguminous shrub is used for animal fodder instead of mulch. In mountainous regions such as the Philippines, the hedges are planted along the contour, and they serve the additional function of preventing erosion. In Peru, contour hedging conserved 287 mm water and 73 t/ha soil annually, but in only 3 out of 15 crops was yield of rice and cowpea increased. Lack of response was partly due to the fact that 22 percent of the land was lost to the hedgerows, and partly that a longer time span may be required to realize the benefits of soil conservation (Alegre and Rao 1996).

Because of the arrangement of plants in rows, alley cropping is amenable to mechanical cultivation, an important factor when labor costs are high. Alley cropping could have potential in the temperate zone where restoration of degraded soils begins to take on equal importance with high production. Alley cropping also could be useful for organically certified gardeners, because alley cropping is more adaptable to machine cultivation than other types of organic production.

Fallowing for Soil Enrichment

Enrichment of soil during fallowing has been practiced throughout the world for thousands of years. After several years of cultivation of crops, the soil becomes depleted of nutrients. During a fallow period, the trees increase nutrient stocks in several ways. Leguminous trees fix nitrogen from the atmosphere; exudates from the roots of certain wild species mobilizes phosphorus that is bound to iron compounds in the soil; deep roots of some species extract calcium and potassium from deep soil horizons (Jordan 1985). The length of fallow required before cropping is again feasible depends upon the characteristics of the soil. In richer soils (such as volcanically derived Andosols), fallows can be less than five years, whereas in highly leached and weathered soils of the central Amazon, several decades may pass before the soil is restored enough to produce a new crop.

During shifting cultivation, indigenous farmers will abandon plots where the soil is depleted and move on to new areas. Often, they will return after fertility is restored. In cultures where land is owned, farmers usually cannot afford to leave their land in fallow for extended periods of time. There, *green manuring* has sometimes been used to restore fertility to the soil more quickly. Green manure refers to leguminous herbs such as clover and alfalfa that are interplanted, or cropped in rotation with grain plants. Leguminous plants are used because they are more efficient at enriching the soil fertility, especially for nitrogen.

The technique of enriching soil that has been depleted by annual harvest of grain is not new. Clover became a major crop in northern

Europe in the eighteenth and nineteenth centuries because of its ability to enrich the soil. Other leguminous crops such as peas, beans, and vetches had traditionally been included in a crop rotation with a grain, often oats. The ability of clover to enrich the soil, however, was much greater. Clover was sown together with the grain crop. By the time the grain was harvested, a growing field of clover emerged that had time to thicken during the autumn. The following year, the clover was usually cut for hay, and following that, it was grazed for several years (Kjaergaard 1995). In Mediterranean climates where clover and alfalfa do poorly, an annual called medic is frequently grown as a green manure (Chatterton and Chatterton 1996).

Although green manures enrich the soil, they have relatively low economic value compared to grain crops. Most are annuals that do not survive well during dry periods. Further, when they were used in rotation, they occupied land that could otherwise be used for the grains. With the advent of cheap and readily available inorganic fertilizers, it became more economically profitable for farmers to use these instead to supply the nutrient demands of their grain crops. To a large extent, the use of green manures was then discontinued in temperate zone countries. In many regions of the tropics, however, where inorganic fertilizer is expensive and a wide variety perennial legumes are available for fallow, green manure is still a good option.

MIXED SPECIES FOREST PLANTATIONS

In conventional even-aged monoculture plantations, trees are planted close together to lessen the problem of weeds for the first few years. Dense planting also forces trees to grow quickly in height rather than in diameter, resulting in a more desirable cylindrical shape of the trees. The disadvantage of close spacing comes after a few years, when the trees get large enough to compete with each other. Competition results in decreased growth, and eventually stagnation (Walker 1984, Nilsson 1993). When competition causes growth of trees to stagnate, the stands must be thinned by cutting out approximately every other tree. Since the small trees usually have little or no market value and thinning is labor intensive, the process can be economically wasteful.

To minimize the problem of competition in forest stands and to foster ecological complementarity, I have begun a series of experiments on the Georgia Piedmont with mixed species plantations of trees. One species I use is *Paulownia fortunei*, a tree that can yield logs valuable for furniture and veneer within twenty years. To obtain a good market price, it is important that the bole of paulownia is kept straight and clear of branches. Interplanted with paulownia are a number of other species with a round shape and full lower branches that compliment paulownia ecologically. They provide side shading that kills the lower lateral branches of paulownia and forces the paulownia to grow straight and tall. One of the complimentary species is poplar, valuable for pulp. Others include species valuable as Christmas trees normally harvested after four to eight years — at about the time that a monoculture of paulownia would have to be thinned. Also interplanted are oaks, which are released when the paulownia is harvested. The combinations of paulownia and other species are planted in lines far enough apart to allow harvesting without damaging the remaining trees.

Economic Complementarity

Often individual landowners hesitate to plant high-value species such as white oak because of the long time interval between planting and economic payoff at harvest time. I am also addressing this problem in my experimental plantations in Georgia. Between the poplar, which can be harvested for pulp at about six years and paulownia, which can be harvested at about twenty, is the slower growing white oak, which takes forty or more years to reach a harvestable size. The harvest of poplar "releases" the paulownia, that is, it gives the paulownia more space to grow. When the paulownia is harvested at fifteen to twenty years, the white oak is released. The time interval until the oak can be harvested is then reasonable from an economic perspective.

Complementarity is also being tried in high-value, long-rotation plantations in the tropics. In Thailand, teak was once an economically important export species, but importance has declined due to overcutting natural forests (Gajaseni and Jordan 1990). To rejuvenate the crop, vast areas in Northern Thailand were planted to teak in the 1960s. A problem with teak plantations is the sixty-year time interval required before a return on the investment. There have been two different approaches toward solving the economic problems that result from a long rotation. One is to allow local farmers to plant grain and subsistence crops between the newly planted seedlings of teak in the *taungya* system (Gajaseni 1992). This results in a benefit for the teak in that the weeding carried out by the farmer also benefits the teak. It also provides a local income during the first years of the plantation.

A second approach to providing economic income during the course of a teak rotation is to interplant with tree species giving a yield during the first decades of the plantation (Gajaseni and Jordan 1992). Tamarind (*Tamarindus indica*) a tree with a round, ball-shaped canopy that complements the long straight trunk of teak, has been tried. It produces a popular fruit used in the preparation of Worcestershire sauce. Tamarind begins bearing fruit in less than five years and produces for five to fifteen years. After fifteen to twenty years, it will be naturally shaded out by the teak, and thinning may not be necessary.

Complementarity in the Humid Tropics

Tropical rain forests are one of the most diverse communities in the world. Because of the rapid rate at which they are disappearing, reforestation efforts are increasing. However, most of the reforestation is with monocultures of pine, eucalyptus, and teak (Evans 1992). Because of the problems of intraspecies competition and of the rapid spread of pests in monocultures, scientists have begun experimenting with plantations of mixed native species. In Costa Rica, Montagnini *et al.* (1995a) found that mixed plantations of native species had greater growth and lower pest damage than pure stands for three of the twelve species tested. One of the reasons for greater growth in mixed stands could be that species compliment each other in their nutrient cycling strategies (Montagnini and Sancho 1994).

To compare the growth of monocultures in the humid tropics with plantations of mixed native species, Batmanian (1990) established replicated plots of *Eucalyptus torelliana*, *Acacia mangium*, and a mixture of six native species. For the first few years, the eucalyptus and acacia grew much faster than the native species. This may have been because the populations used commercially have been selected to respond better to plantation conditions — direct sunlight and fertilization. Only one native species, *Schizolobium amazonicum*, an early successional legume, grew well during the first few years. After about four years, however, the acacia stands were infected with a parasite from the mistletoe family that severely damaged or killed all the trees within one year. The eucalyptus stands stagnated because they were not thinned. In contrast, as the stands of mixed native species began to get larger, their rate of growth increased. The increase was due in part to decreased competition from herbaceous weeds and grasses, and to an increasingly efficient use of light and nutrients. A similar result was found in a comparison of pine and native species in Puerto Rico (Jordan and Farnworth 1982). Initially, the growth of pine was greater, but eventually, the native species were more productive.

Many tropical rain forest species are adapted to the partial shade and humid conditions of the undisturbed forest. When these species are planted in the open sun and bare soil, they often do poorly. Young seedlings and saplings of many rain forest trees do much better when they establish under a "nurse" species that lessens the severity of the microclimate around the seedlings, partially suppresses weeds, and provides leaf litter that increases the soil organic matter (Mesquita 1995). Good nurse species often will be early successional species that can establish themselves in harsh microclimate and disturbed soils. An economic problem with early successional species is that they often have little economic value. Some species such as *Cecropia peltata*, however, have value as pulp.

Structural Diversification

Most plantations are not only monocultures, but they are also even-aged, because it is cheaper and easier to plant, harvest, and market even-aged stands. Some economically important trees such as Douglas fir grow well in monocultures, and there might be little advantage in attempting to diversify a stand with more species. However, even-aged monocultures of trees are structurally uniform, which has its disadvantages. There is intense intraspecies competition when all the trees are the same size. Competition can be decreased by managing for structurally diverse stands. Once an uneven-aged stand is established, it can be maintained by periodically harvesting only the largest trees. This opens up the stand and allows more sunlight for the smaller trees, saplings, and seedlings.

Another disadvantage of even-aged monocultures is that they are relatively susceptible to disease and fire. For example, pine bark beetle in loblolly pine spreads most rapidly in mature stands. The beetle spreads rapidly when old trees are closely adjacent. When trees are planted in small blocks of the same age, the spread of the beetle is inhibited.

Another example is from the spruce-fir forests of Canada. The spruce budworm is an important pest in mature stands and can rapidly infect large areas, killing trees and making the remainder highly susceptible to fire. By managing the stands in blocks of various ages, the problem can be lessened. In addition, such a strategy is advantageous to mill operators who prefer a steady stream of pulp supply to periods of excess and shortage (Erdle and Baskerville 1986).

MANAGEMENT FOR FACILITATION AND COMPLEMENTARITY

In the past, agronomists and foresters have tried to increase production by trying to diminish or eliminate competition between crop species and weeds, or between individuals within a crop. Mechanical and chemical methods have often proved to be the most effective control for interspecies (weed) competition, and adjustment of spacing for intraspecies (within-crop) competition. Now, as we begin to see the environmental side effects of over emphasis on mechanical and chemical control, scientists have begun to look at alternative methods of weed control, including innovative systems of mulching, intercropping, and fallowing. As we begin to look at the costs of monocultures in terms of reduced efficiency at the community and ecosystem level, mixed plantations and cropping systems of species that compliment each other are appearing more attractive, both economically and ecologically.

CHAPTER 6
INTERACTIONS IN THE SOIL

FUNCTIONS OF SOIL ORGANISMS

In this chapter, we look at the interactions of organisms in the soil, with a view towards how these functions can be used to replace petroleum based subsidies in farm, plantation, and grazing systems. The functional mechanisms that contribute to sustainability in natural systems, are often not immediately obvious. Sometimes, it is difficult to determine if interactions are taking place — and if they are, how they benefit productivity and sustainability. It is especially difficult to analyze interactions in the soil, since taking measurements often disrupts or destroys soil structure. Through laboratory analysis, microbiologists have succeeded in uncovering some of the negative interactions between species in the soil, such as the production of antibiotics by fungi. However, positive interactions that benefit the sustainability of ecosystems often become obvious only when they are studied in the context of an entire ecosystem.

A convenient way to untangle interactions in ecosystems is to follow the flow of carbon with a radioactive tag — from the atmosphere into the leaves of plants, through the herbivores and then the predators, into the soil and the decomposers that live there, and finally back into the atmosphere (Parton *et al.* 1996). Another way to analyze the interconnectedness in an ecosystem is by looking at the flow of energy. It is energy that fuels the interactions that provide both productivity and sustainability.

Energy Flow Into and Through the Soil

For all organisms, the ultimate source of energy is the sun. Plants use solar energy to convert carbon dioxide in the atmosphere into energy-rich carbohydrates. Some carbohydrates are used by the plant while the rest are passed on to herbivores — that is, animals (including insects) that eat plants or plant parts. Some of the energy embedded in herbivores is passed up the food chain and used by predators.

When a plant dies or sheds its leaves and roots, or when an herbivore or predator dies, the carbon compounds that make up the plant or animal become the source of energy for decomposer organisms such as bacteria, fungi, insects, and worms living in the soil. They are called decomposers because when they ingest the complex compounds, they break them down into simpler ones such as sugars. Ultimately, these decomposers respire the sugars, thereby returning carbon to the atmosphere as carbon dioxide.

When a natural ecosystem is cut down and replaced by conventional agriculture, plantations, or pasture, there often is little further input of carbon compounds into the soil. The energy supply for the soil organisms is, to a large extent, eliminated. As a consequence, the functions previously performed by soil organisms must now be done artificially through chemical fertilizers, pesticides and mechanical cultivation. The substitution of natural functions by artificial subsidies is a basic cause of unsustainability in resource production systems.

Nutrient Recycling

There is always a certain amount of nutrient loss from natural ecosystems. Calcium, potassium, and magnesium are leached. Nitrogen and sulfur are volatilized. However, the rates of such losses are usually low, and nutrients lost through natural processes can be compensated for by natural inputs. Many nutrients are imported into ecosystems while dissolved in rainfall or adhered to fine particulate aerosols that are blown in from other regions. In natural ecosystems, the ultimate source of nutrients such as calcium, potassium, and phosphorus is the bedrock that underlies the soil. Nitrogen enters the soil from the air as a result of bacterial activity, and atmospheric sulfur is taken up directly by plants and soil organisms.

The rate at which elements enter an ecosystem from the soil and atmosphere is very slow when compared to the rate at which plants must take up nutrients to grow and reproduce. Most of the nutrition for plants in natural ecosystems comes from nutrients that are recycled within the ecosystem (Jordan 1985). Recycling of nutrients in an ecosystem occurs when carbon and its embodied energy is passed through the food chain from plants to animals to decomposers (Odum and Odum 1981). The flow of most nutrient elements parallels the flow of carbon, with the exception of carbon movement from the atmosphere during photosynthesis and carbon release back to the atmosphere during respiration.

The nutrient elements join the flow of carbon through the ecosystem shortly after carbon dioxide in the air is transformed into sugars in leaves. As the sugars are transported throughout the plant, other nutrients are added as more complex compounds such as proteins are synthesized. The flow of nutrients departs from the flow of carbon during the final stages of decomposition when the carbon of dying microbes is respired into the atmosphere. At that time, the nutrients are released in soluble form and are taken up by the roots of plants.

Flow of energy, carbon, and nutrients through the soil begins when freshly fallen litter and other detritus on the soil surface is colonized by bacteria and fungi. As they respire the carbon compounds, they leave the remaining organic matter with a higher concentration of nitrogen and phosphorus, making it more palatable to a wide variety of insects, flagellates, and worms. The detritus eaters are preyed upon by arthropods, including mites and spiders, worms (such as nematodes), and amoebae. Some of the predators in the soil also eat bacteria and fungi. At each level of the decomposer food chain, some of the carbon contributes to growth of the organism, while some is respired, some is egested, and some is passed along to predators or decomposers. As soon as carbon in an organism becomes available, it is pounced upon by other organisms that utilize part of it and then pass the rest along until all of the carbon's energy is dissipated through respiration, and it is released as carbon dioxide (Coleman and Crossley 1996).

As carbon moves in these mini-cycles, the nutrients that are a part of the carbon-based compounds also are cycled. At each step of the way, some of the carbon is respired away. In contrast, the nutrient elements are passed along with negligible loss to the next participant in the food chains. There is little opportunity for leaching of calcium and potassium by the water that percolates down into the lower soil. There is also little nitrogen in the form of ammonia that can be volatilized or in the form of nitrates that can be leached or denitrified. Very little phosphorus is lost through fixation by aluminum in the mineral soil, because most of it is taken up by other organisms before it ever comes in contact with the mineral soil (Coleman *et al.* 1994).

These mini-cycles of nutrients within the soil however, are not completely efficient. There is always some gradual leakage of nutrients, and it is this leakage that is available for uptake by roots of plants. Because the rate of leakage is slow, there is usually ample opportunity for roots to absorb the nutrients before they are lost.

Major losses of nutrients from ecosystems generally occur when the nutrients are not incorporated in the food chains of the soil. This occurs when the input of organic matter ceases, as when deforestation occurs and an area is converted to agriculture. As the community of soil organisms begins to die, nutrients gradually begin to be lost from the remaining organic matter. As the nutrients then come in contact with mineral soil, electrostatic exchange on the surface of clay particles becomes the principal mechanism through which they are held in the soil. When nutrients are held only by exchange, they are readily lost. This loss occurs through the replacement by hydrogen and anions from organic acids in the soil. Once the nutrients are dissolved in the soil water, they are highly susceptible to leaching, volatilization, and fixation (Brady 1974).

An important reason that natural systems are sustainable systems is the nutrient recycling function of the soil organisms. In the Amazon region of Venezuela, we determined the efficiency of nutrient recycling in a rain forest where a layer of decomposing soil organic matter covers the soil (Stark and Jordan 1978). Over a period of two months, 99 percent of the radioactive phosphorus and calcium applied to the surface was recycled, and less than one percent was lost through leaching.

In agricultural systems maintained by conventional plowing and tilling, the food chains that maintain the cycles of nutrients are destroyed. Nutrients are quickly lost and must be replaced by inorganic fertilizers. The food chain can be kept intact if the field is mulched, but if plowing is the primary method of soil preparation, the soil food chains are quickly disrupted. Research into sustainable agriculture must focus on methods of keeping the food chains within the soil intact (Gupta 1994).

Mycorrhizae

Because of the tight food chains within the soil of ecosystems that are high in soil organic matter, it can be difficult for higher plants to acquire the nutrient elements they need. Roots of plants often cannot grow fast enough to outcompete the bacteria and fungi when a nutrient element becomes "available" — that is, when the element is released in soluble form as a result of either egestion or the death and decomposition of a microorganism. Higher plants, however, are competitive for nutrients as a result of a symbiotic relationship with a type of fungi called *mycorrhizae* that grows in intimate contact with the roots of the higher plants. The branches, or *hyphae* of mycorrhizae, penetrate much of the space within the decomposing organic matter, and are often able to outcompete other fungi or bacteria when a nutrient molecule becomes available. The mycorrhizae then pass along the nutrients to the plant, and in exchange, obtain energy-rich carbohydrates for metabolism from the higher plant (Swift *et al.* 1979). While many crop plants have mycorrhizal associations, the fungi may not perform an essential function in croplands, where fertilizers are the main source of nutrients and fewer soil organisms will compete for that fertilizer (Ruehle and Marx 1979).

Functions Resulting from Diversity

Demand by plants for nutrients is spread throughout the entire year. Some species need nutrients early in the growing season, while others need them later. Different phases of growth, such as vegetative growth and fruiting require different combinations of nutrients. For plants, balancing this diversification of need is a diversification of nutrient supply. The flow of energy and nutrients through ecological food chains is not steady throughout the year, but varies according to season, climate, and pressure of herbivores such as insects (Swift and Anderson 1993).

The nutrient supply depends, in part, upon the characteristics of the litter on or near the soil surface (Montagnini *et al.* 1995b). Faunal remains are high in nitrogen and decompose very quickly. Leaves of leguminous plants often decompose more slowly than animals (except for bones), but usually more rapidly than other plants. Tree leaves that are high in secondary plant products, such as tannins and lignins, decompose slowly. Almost every species of plant and animal has unique characteristics which are important in determining its rate of decomposition. As a result of the diverse supplies and demands for nutrients, natural ecosystems usually exhibit a tight synchrony of plant and microbial activity (Woomer and Swift 1994).

Diversity also is important in disease suppression. Root pathogens are usually specific to a particular type or species of plant. In wheat and barley cropping systems where root disease can be a problem, control is sometimes obtained through rotation of crops. Because of the diversity in natural systems, it is more difficult for a root pathogen in one individual to spread to another than in cropped systems. Root diseases also are less of a problem in the soil of natural ecosystems because of the presence of parasites of fungi, competitors and predators of disease organisms, and antibiotic-producing microorganisms known collectively as *resident antagonists* (Cook 1994).

Soil Structure

The primary components of soil are fine mineral particles released from weathered rock. These components vary in size from fine clay to coarse sand. Pure sand is not a good medium for the roots of plants, because water quickly drains through it, carrying away nutrient elements. Rapid percolation of water also causes the sand to dry out quickly after rainfall, creating a water shortage for the plants. Organic matter, however, transforms inhospitable sand into soil favorable for plant growth. Transformation of sands can be seen on dunes along the beach. Through random acts of nature, a piece of organic debris — such as seaweed, or a log — lodges at the top of the dunes, and seeds of dunegrass soon germinate in this more hospitable environment. The grass quickly forms underground rhizomes that spread to form a network that traps nutrients. On the back of the dune, organic matter begins to accumulate, and soon trees and shrubs are growing. If no big storms occur that could wash away the trees, a forest will eventually evolve.

Pure clay is not a good medium either for the roots of plants. Pure clay is dense and compact, and water tends to run off the surface rather than infiltrate. Further, it is difficult for roots to grow through clay. However, when organic matter is added to clay, it is transformed into a medium much more hospitable for the roots of plants. Organic compounds of microbial origin bind clay particles together to form aggregates stable enough to withstand the impact of falling rain. Worms and some insects that feed upon organic matter also contribute to the porosity of the soil. A soil composed of clay particles supports plant growth, because the soil organic matter and the activities of soil organisms render the soil permeable to water, and thus permeable to roots.

The structure of soils that allows them to be productive depends upon a continuous flow of energy in the form of carbon compounds through the soil organisms (Cheshire 1985, Parsons 1985). When the flow of carbon is halted as a result of clearing, plowing, and cultivating, soil structure breaks down. Further production is then possible only through intensive management including plowing, disking, and fertilizing.

Protection, Microclimate, and Disease Suppression

In natural ecosystems, there is often a layer of organic matter lying on top of the soil. On the top most layer are relatively undecomposed leaves. Beneath that are remnants of leaves partially eaten by insects or decomposed by microorganisms. These layers are important in maintaining soil conditions that are beneficial for plants and soil organisms. Surface layers of organic matter reduce the energy of drops in rainstorms, preventing the breakdown of soil aggregates that allow the water to percolate into the soil. When there is a layer of litter, surface runoff erosion is rare.

The litter layer also acts as insulation. It narrows the temperature extremes within the mineral soil and usually improves conditions for seed germination and root growth. It also inhibits evaporation from the soil surface. While insulation is almost always beneficial in tropical and some temperate ecosystems, it can be detrimental in high-latitude ecosystems where low temperatures inhibit decomposition, plant germination and growth.

Soil organisms also are important for sustainability in that they can reduce crop losses to insects and diseases. For example, egg-laying by the European corn borer was lower in soils with a history of organic farming (Phelan *et al.* 1995).

CHANGES IN THE SOIL SYSTEM FOLLOWING CULTIVATION

Destruction of Food Chains and Energy Flow

A natural system (such as a tropical rain forest) is a sustainable system, because the functions of sustainability are performed by the living soil system. Sufficient energy to maintain this soil system is supplied by the litter input from the natural forest. When a natural ecosystem is clear-cut for logging or converted to a cropped system, the energy supply for the soil organisms is reduced or eliminated. As a result, the functions for sustainability are also reduced or eliminated (Baumgartner and Kirchner 1980).

When a forest is cut down, there can be enough organic matter remaining on the soil surface to keep the soil system intact and functioning for several years before lack of carbon input causes the system to break down. This remaining carbon is the basis for shifting agriculture which is still practiced in some areas of the tropics. Farmers practicing shifting cultivation often burn the woody debris after cutting the forest, but the fire usually burns only the fine twigs and branches. It does little damage to the soil organic matter and leaves the trunks intact (Jordan 1989). It is when the soil organic matter disappears that the nutrient recycling capacity of the soil is diminished, and the farmer must move on and clear another plot.

In "modern" farming, loss of soil organic matter during cultivation depends upon the intensity of cultivation (Grace *et al.* 1994). Because annual crops such as corn and cotton require frequent intensive disturbances to the soil, these crops do the greatest damage to the soil system. As a result of the churning of the soil by plows and cultivators, very little organic matter may remain in the soil. Consequently, the community of soil organisms is severely damaged.

Reduction in Nutrient Retention

When the community of soil organisms is destroyed through agricultural or forestry practices, nutrients are not immediately lost, but rather are adsorbed on the surface of clay by electrostatic forces in a process called *exchange* (Tan 1993). The nutrient-retaining power due to exchange is relatively weak. As a result, nutrients can be readily taken from the clay surfaces either by roots or by mycorrhizae (in association with roots). In a cropping system, however, only a portion of the nutrients in the soil are actually taken up by the crop plants. This is because the living roots of the crops, especially in a newly planted crop, only occupy a small proportion of the soil mass. Some of the nutrients remaining in the soil are taken up by weeds. If the weeds are controlled through cultivation or herbicide treatment, the nutrients again return to the mineral soil. Nutrients exchanged in the mineral soil that are not taken up by weeds or crops are susceptible to rapid loss.

Some nutrient elements, such as calcium and potassium, have a positive electrical charge, and are held on the surface of clay particles by negative charges. Potassium is one of the first elements to be lost once food chains in the soil disappear, because it is held to the mineral soil by only a single charge. During the first heavy rain, potassium exchanged on the clay will be replaced by hydrogen dissolved in the soil water, and will be lost to the ecosystem as the water percolates down through the soil. Calcium and magnesium, held by double charges, are lost through the same mechanism, although somewhat more slowly.

Other nutrients, such as nitrate-nitrogen and sulfate-sulfur have a negative charge and are held at sites of positive charge. Like potassium and calcium, nitrogen and sulfur can be taken up by the roots of crops and weeds, but if uptake does not occur rapidly, the nutrients can be leached away by percolating ground water or volatilized through bacterial activity.

In soils that are high in iron and aluminum, phosphate quickly becomes unavailable to plants when it comes in contact with the mineral soil, where it forms iron and aluminum phosphate. Phosphorus in these states is highly insoluble. Because soils high in iron and aluminum are ubiquitous in the tropics, phosphorus deficiency is common in many cultivated tropical ecosystems.

Creation of Bottlenecks

Diversity of an ecosystem contributes to the stability and efficiency of nutrient recycling. Conversion of natural ecosystems to single-species plantations of trees destroys much of the diversity of the system, and thereby eliminates the contribution to sustainability offered by diversity. In simple ecosystems with few species, bottlenecks may occur in ecosystems because of insufficient microbial populations or microbial activity to keep the carbon and nutrients moving through the food chains. Undecomposed litter may accumulate on the forest floor, depriving higher plants of a source of nutrients. With a greater number of species present in the soil, the probability of irregularities in the flow of nutrients and energy through the soil decreases.

Influence on Microclimate

As a result of the disappearance of litter and humus on the soil surface, the soil microclimate becomes more severe. While litter and humus reflect some of the sun's energy and have a low capacity to retain heat, mineral soil absorbs almost all of the energy incident upon it and converts most of it to heat. Consequently, evaporation is much higher from bare mineral soil than from litter-covered soil, and soil without cover dries out more quickly. Hotter, drier soil is a less hospitable environment for plants, and germinating seeds are especially sensitive to heat and drought (Barrett *et al.* 1990). High soil temperatures also increase rates of microbial activity, thereby stimulating breakdown of soil organic matter in the lower soil horizons.

Changes in Soil Structure

Organic compounds and humus are the materials that bind soil together and give it structure. When organic matter input to the soil ceases, soil structure breaks down (Coleman *et al.* 1994). What is left in clay soils is a very dense, highly impermeable mass. Rainwater tends to run off the surface and cause erosion, instead of infiltrating into the soil. It is very difficult for roots to permeate. In sandy soils, loss of organic matter can result in increased rates of percolation, with the result that the soils become droughty.

Decrease in Suppression of Disease

The disappearance of a detrital food chain in the soil may result in the disappearance of disease-suppressing organisms that are part of the living soil community (Neate 1994).

HOW WE COMPENSATE FOR THE LOSS OF NATURE'S SERVICES

Petroleum Subsidies

When the farmer, forester, or range manager completely clears away the natural ecosystem in order to plant annual crops, plantation seedlings, or pasture grass, he must provide substitutes for the services of nature that have been destroyed.

Losses of natural functions and the artificial subsidies that replace them include the following:

- The loss of nutrient recycling functions, which is replaced by fertilization.

- The loss of structure in soils, which is compensated for by plowing, disking and harrowing.

- The loss of soil cover, which decreases evaporation from the soil. It is compensated for by irrigation.

- The loss of disease control functions, weed control functions, and herbivore control functions, which are compensated for by chemical control.

- The loss of genetic diversity, which is compensated for by genetic engineering.

All of these activities involve energy supplements, either directly or indirectly. In mechanical plowing, petroleum energy is used directly to power the tractor that breaks the soil. It is also used to power the pumps that irrigate the fields. Petroleum energy is used to synthesize and purify inorganic fertilizers, and as a base for pesticides and herbicides. Genetic engineering is dependent upon an energy-intensive infrastructure. For these reasons, agriculture that is dependent upon such techniques has been called *energy-intensive agriculture* (National Research Council 1989). Energy intensive agriculture is inefficient agriculture, because functions that could be performed by energy from the sun instead are performed through the use of petroleum energy (Pimental *et al.* 1973).

Alternatives

In natural ecosystems, the native species supply the organic matter (mainly leaf-fall and root exudates) that nurture the below-ground ecosystem. In agricultural systems managed for annuals or other relatively short-lived crops, organic matter can be imported into the system from elsewhere. An alternative to energy intensive agriculture, forestry, and range management is through management of soil organic matter.

Organic Farming

In *organic farming*, litter and mulch are typically produced in one area of the farm and transported to, and spread upon, the economically important crop. There are almost as many types of mulch and mulching systems as there are farmers. A wide variety of materials from the farm and forest can be used to build up or maintain the soil organic matter in fields cropped for annuals. Wood chips, bark, leaves, pine needles, peanut shells, pecan husks, hay, straw, and leguminous herbs (such as alfalfa) are common examples. The ambitious home gardener creates a compost from table scraps and grass cuttings (Martin and Gershuny 1992). Certain species such as pine cannot be used as mulch until the resins and tannins in the bark and wood decompose, as they can inhibit growth of crop plants.

Depending upon the purpose of the mulch, it may have to be mixed with manure. If the soil is already in good condition and the purpose of the mulch is only to improve the microclimate and to maintain the organic matter content of the soil, the mulch can be applied directly. If nutrient elements are critical, addition of raw mulch alone is insufficient. Grain crops need an input of nutrients greater than can be supplied through the decomposition of raw organic mulch. The ratio of carbon to nitrogen and other nutrients in raw hay, wood chips, nutshells and the like is high — 30/1 or more. As a result, decomposition of the mulch is slow, and release of the nutrients that are present in the mulch is relatively slow.

There are several solutions. The raw mulch can be mixed with manure, which is relatively rich in nutrients. It also can be composted — that is, let alone so that bacteria and fungi respire away the carbon, leaving the remaining material enriched in nutrients. Rates of composting can be increased by the addition of leaves, manure, and other material relatively high in nutrients, or by small amounts of inorganic fertilizer. The nutrients stimulate the growth of the microorganisms that break down the complex carbohydrates which compose the wood chips, straw, and peanut shells. Adjustment of acidity by the addition of lime can also increase the rate of decomposition, depending on the nature of the decomposers.

The major drawback to this type of organic farming is the effort necessary to move the organic matter from where it is produced to where it is needed. It can be labor-intensive, and thus expensive. When it is mechanized, special types of equipment are often necessary.

Crop Rotations

Sometimes a farmer will plant an annual grain one year and then either plant a leguminous crop the next, or a cover crop during the winter. The primary purpose of the alternate crop is to provide an organic matter input into the soil and to inhibit the buildup of pest populations. The advantage of a leguminous crop rotation is that the organic matter does not have to be moved, but can be plowed directly into the soil (National Research Council 1989). In some regions, governmental soil conservation programs will pay farmers to keep their field out of annual grains and in systems which restore soil organic matter.

The paddy rice system that is traditional in parts of Asia takes advantage of organic matter that is produced exactly where it is needed. Paddy fields are flooded, either naturally or through regulation of a series of dikes, and a nitrogen-fixing fern called *azolla* produces organic matter within the paddies. Then the fields are drained, and the nutrients and organic matter from the azolla are available to the rice (Soermarwoto 1977).

Conservation Tillage

There are three basic reasons that the farmer plows the soil. One is to uproot weeds that compete with the crop. The second is to ensure that seeds of the grain crop come in contact with mineral soil, where soil moisture is adequate to promote germination. The third is to loosen the soil so that the new roots can grow more readily.

Plowing and mechanical cultivating are agricultural techniques that devastate the soil. Erosion is one problem, and plowing and cultivating along contours has been promoted as a solution to the problem. Erosion, however, is not the only negative effect of plowing. Plowing turns up the mineral soil and exposes bacteria to the air. As a result, their activity increases, and the organic matter residue in the soil quickly disappears. Another problem is the compaction of soil due to mechanical cultivation.

In recent years, there has been an increasing adoption of no-till and minimum-till systems of cultivation. These systems facilitate the buildup of organic matter in the soil, and reduce erosion and compaction. In no-till planting, the first step is to knock down any standing residue of crops and weeds. Then the no-till planter is drawn over the field. It has discs that make a slit just wide enough to accommodate a seed. A seed is dropped into the furrow immediately behind the discs, and then the slit is tamped closed by a wheel. When no-till planting is used continually, the soil increases in organic matter and decreases in density, thereby improving it for crops. In minimum-till systems, the soil may be raked or treated with circular blades, but it is not turned over with a plow.

The problem with no-till and minimum-till planting is weed control. Frequently, herbicides are used to kill weeds immediately before the crop is planted. It is not really necessary, however, to completely eliminate all the weeds. All that is necessary is to ensure that the crop plants get a head-start on the weeds, or that the weeds do not outcompete the crops. One approach is to plant the crops close enough together so that they shade out the weeds.

Another technique that I have used on my farm in Georgia is to plant clover as a winter crop. As mentioned in Chapter Five, it flowers in late April and dies back in May, forming a thick mulch in which there are almost no weeds. Then, we plant sorghum with a no-till planter, and by the time weeds finally become established, they are no longer a threat to the sorghum.

Agroforestry

In agroforestry systems, trees and annual crops are intermingled. Leaf litter supplies the energy required by the soil community (Young 1989). In some agroforestry systems, trees and annual crops are mixed in one field. No effort is required on the part of the farmer to import the organic matter. Litter from the trees falls directly on the soil surrounding the crops. In the wet tropics, there can be an almost continuous input of leaf litter from trees in home gardens.

In temperate regions, leaf-fall usually occurs in a single pulse at the end of the growing season. The supply of leaf litter is not well synchronized with the demand of crops that may be interplanted. Alley cropping is a potential solution to this problem. The hedges are mechanically pruned several times during the growing season, and the leaves and small branches fall into the alley and serve as a source of soil organic matter. Alley cropping has the potential to combine the advantages of organic subsidies to the soil with the economic efficiency of machine cultivation in countries where labor is expensive. It can be a profitable approach to organic farming (Kang *et al.* 1990).

Sustainable Forestry

There are a wide variety of logging practices which vary in their effect on forest soils. Clear-cut logging exposes the soil to the full impact of rain, with the resultant problems of erosion. Selective logging, where only a small proportion of the standing crop of trees are taken out, may have a minimal impact if the logging is done carefully. The remaining trees are capable of providing enough litter to keep the soil protected and intact, with only a slight reduction in soil quality and nutrient recycling functions. Logging practices which leave dead trees in the forest also are more beneficial than salvage logging, which cleans the forest floor. Dead and decaying trees in a forest are an important source of energy for nutrient recycling and ecosystem sustainability (Harmon *et al.* 1990).

Other practices which increase the sustainability of harvested forests include: building roads along the contour; extending skid trails up and downslope from the roads in a herringbone pattern; using metal culverts where logging roads must cross streams; directional felling, where trees are felled to reduce damage to trees targeted for future harvest; and in tropical forests where vines commonly link canopies of adjacent trees, cutting vines two years before harvest, so that trees that are felled do not uproot neighboring trees (Gerwing *et al.* 1996).

In plantation forestry, mixed-species plantations can be more beneficial for maintenance of soil fertility than monocultures. Each species combines nutrients in a unique way. A variety of trees is better able to maintain a flow of carbon through the soil food chains because of the more diversified supply of nutrient elements that become available to the decomposers. With good (greater than two percent) levels of organic matter in the soil, other nutrients become more readily available. Microbes in soil organic matter are important in sulfur retention in the soil (Scott 1985); thus, management techniques which increase soil organic matter should help satisfy sulfur requirements of crops. When phosphorus is limited in acid soils due to fixation by iron and aluminum, the problem can be remedied by species (such as *Gliricidia sepium*, a tropical tree) which exude compounds into the soil that mobilize the bound phosphorus (Jordan 1995).

Economic Considerations

The decision of farmers to subsidize crop production with petroleum-based subsidies (instead of natural-based subsidies) is often economic. Petroleum-based subsidies are cheaper in the short term, and the competitive economic system forces farmers to maximize short-term profits. Unfortunately, maximization of short-term profits comes at the expense of long-term sustainability, which is dependent upon soil organic matter.

Soil organic matter is *capital* in the same sense as a factory is capital (Folke *et al.* 1994). The factory owner, however, is allowed to calculate depreciation of his factory on his income tax. The owner of an oil well is allowed to calculate depletion of his stock of petroleum. In contrast, the farmer is *not* allowed to depreciate the value of his field as the stock of organic matter decomposes, as the system does not recognize the value of natural capital. The economic system gives the farmer little incentive to farm sustainably.

CONCLUSION

Soil organic matter can be maintained in a number of ways, some of which have been practiced since the beginning of agriculture. What is new is our understanding of how soil organic matter sustains production functions. This understanding gives a scientific underpinning to the movement toward a more sustainable agriculture.

CHAPTER 7

PEST AND DISEASE INTERACTIONS

The complex and diverse structures and functions of nature help to prevent outbreaks of pest populations and the spread of disease in unmanaged ecosystems. The key to increasing the resistance of managed production systems (cropland, plantations, and pastures) against insects and disease is to incorporate pest-inhibiting functions of natural systems into the economic systems. In this chapter, we discuss the mechanisms within natural ecosystems that contribute to inhibiting the spread of pests and diseases.

PESTS

There are two major hypotheses (Altieri and Liebman 1986) to explain why pests are often less of a problem in complex and diverse crop ecosystems than they are in simple ones. The *Natural Enemy Hypothesis* predicts that there will be more predators that prey on insect pests in complex polycultures than in simple monocultures. One reason for this is that polycultures can provide a more reliable source of food for the predators. Some predators often consume a wide variety of prey, and therefore are more likely to find a continuous source of food in a heterogeneous habitat. Other predators are specialized, but ecosystem complexity also helps maintain their populations. Their populations are less likely to fluctuate widely in a heterogeneous system because its complexity allows the prey to escape local extinction (Root 1973).

The other major hypothesis is the *Resource Concentration Hypothesis* (Root 1973) which predicts that insects are more likely to find and remain on hosts that are growing in monocultures. In complex, diverse systems, pest species with a narrow host range have greater difficulty in locating and remaining upon host plants in small, dispersed patches as compared to large, dense, pure stands.

There are a number of other mechanisms that underlie hypotheses based upon natural enemies and resource concentration (Hasse and Litsinger 1981):

1. In diverse environments, adult predators are more likely to encounter suitable habitat, nectar and pollen sources, thus reducing the probability that they will leave or become locally extinct (Risch 1981).

2. In a complex system, overlapping leaves of various species camouflage the host crop, making it more difficult for the pest to locate the host species that it is seeking. For example, in a system where beans are planted in the stubble of a rice field, the stubble makes bean seedlings more difficult for the beanfly to find.

3. Some types of pests are more attracted to crops with a bare background of soil (as in a weed-free monoculture) than to ones with a green background (as in a polyculture). Aphids and flea beetles are more attracted to crops of cole with a background of bare soil than to ones with a weedy background.

4. Nonhost plants can mask or dilute the attractant stimuli of host plants, leading to a breakdown of orientation, feeding, and reproduction processes. A flea beetle that specializes on plants in the mustard family and that is a pest on cole is reduced in diversified systems (Altieri *et al.* 1990).

5. The odors of certain plants also are repellent chemical stimuli, and act as pest repellents. Grass borders

around bean patches help repel bean leafhoppers. Populations of *Plutella xylostella* are repelled from cabbage/tomato intercrops.

6. The complexity of structure in polycultures can interfere with the population development and survival of pests in a number of ways. *Companion* plants may physically block the dispersal of herbivores across a polyculture, or reduce the statistical probability that an insect entering a field will find an attractive host. Decoy crops, or *trap crops*, are species interplanted with economic crops that cause pests to waste their infection potential. For example, pineapple plantations are sometimes interplanted with tomatoes. The tomatoes attract larvae of harmful nematodes, and then the tomatoes are destroyed before the nematodes can fully develop (Palti 1981). In addition, companion plants can alter the microclimate, rendering conditions less favorable for pest species.

7. Secondary plant chemicals (such as polyphenols, alkaloids, and tannins) often common in wild plants of the tropics have apparently evolved as defenses against herbivory (Jordan 1985). The presence of such plants within cropping systems inhibits pests.

DISEASE

Plant pathogens transmitted by insects — often aphids, leafhoppers and planthoppers — cause significant yield losses in crops (Power 1990). Most of these pathogens are viruses, but various mycoplasmas and bacteria are also transmitted. These pathogens can be persistent, semipersistent, or nonpersistent within the insect vector. Once an insect acquires a persistent pathogen, it can be carried by the insect for long periods up to the lifespan of the vector. Probability will be high that it transmits that pathogen to a host plant. In contrast, a nonpersistent pathogen may be retained in an insect for minutes to hours, and a semipersistent pathogen from hours to days. Unless the insect carrying a nonpersistent pathogen finds a host plant relatively quickly, the disease will not be spread.

Once a plant is infected, a period of time must pass before another insect can pick up the pathogen from that plant and spread it to another plant. This period of time is called the *latent period*. When the latent period is long relative to the life of the plant, probability is low that the disease will spread. The dynamics of disease spread in such a system is called monocyclic, because the source of the disease does not increase within the crop during the course of a single growing season (Thresh 1983). An example is the Fusarium wilt in tomatoes, which results from roots contacting fungal propagules present in the soil from previous crops (Mundt 1990). There is little plant-to-plant movement of the fungus within a growing season.

In contrast, when latent periods are short relative to the life of the plant, new sources of disease may develop within the crop. In this case, diseases may be polycyclic and spread from sources of inoculum that multiply as new plants become infected. Because of their long life span, perennial crops such as grapes or tree crops can be hosts to polycyclic pathogens.

Polycultures and Disease Resistance

Just as polycultures are sometimes more resistant to herbivore damage than are monocultures, polycultures also may be more resistant to diseases that are spread by insects. Disease-spreading insects will be confused, repelled, and interfered with by complex ecosystems just as herbivorous insect pests would be. Aerially borne pathogens will also spread less efficiently throughout diverse and complex ecosystems. For example, powdery mildew, a disease that affects wheat, is spread when spores produced from pustules are aerially dispersed. Although the wheat is an annual, the mildew is polycyclid, and many successive cycles of spore production and infection can occur within one season (Mundt 1990). It would seem then, that interplanting wheat fields with shrub or tree barriers could slow the spread of the disease. In China, fields of winter wheat are often interplanted at 50 meter intervals with rows of Paulownia, but the effect upon the spread of disease has not been studied.

Genetic Diversity and Spread of Disease

Evidence from an analysis of the southern corn leaf blight shows the danger of genetic uniformity. In the early 1950s, a variety of corn was uncovered in Texas with a cytoplasmic factor that causes male sterility. Many farmers adopted this line because male sterility saves time and money by eliminating the need for detasseling. By 1970, 85 percent of all seed corn planted in the United States was Texas cytoplasmic male sterile. This line proved to be extremely susceptible to a strain of blight fungus which causes southern corn leaf blight. Weather conditions, coupled with the hypersusceptibility of the genetically uniform crop, led to a massive outbreak of the disease in 1970. In areas of the Southeast where the disease was most severe, there was 100 percent crop loss, while midwest crop losses often exceeded 50 percent. Because of the shortage of other lines of hybrid seed, it took several years before resistant lines could be widely reintroduced (Real 1996).

Plant Health and Disease Resistance

Maintaining an agricultural crop or a herd of livestock in well-nourished condition is the best defense against disease. Insects are more likely to attack weak and damaged plants than they are healthy and vigorous ones. Diseases are more likely to become established in crops or animals suffering nutrient deficiencies than in those that are well-fertilized or well-fed. In much of modern agriculture, crop nutrition is carried out through the application of inorganic fertilizers, rather than through the promotion of recycling of nutrients through soil organic matter. Inorganic fertilizers work when and where they are available and cheap. In many regions of the world, however, they are expensive and scarce. In such regions, preventative medicine, such as soil rehabilitation, is more economical in the long term than is intensive application of pesticides and fungicides.

INTEGRATED PEST MANAGEMENT

Integrated pest management (IPM) is an important weapon in the arsenal of environmentally minded farmers and entomologists. It is a combination of many different techniques which have the potential to decrease the reliance upon chemical pesticides for insect control. Some of the techniques can be considered to be "working with nature" in the sense that they incorporate an understanding of the functions of natural ecosystems. Others are a continuation of the trend toward increasing dependence upon high technology. We begin with a discussion of the former.

Using Natural Functions

Biological Control Using Natural Enemies

Population outbreaks of pests often occur because there are no predator populations to control them. Importing predator species from regions where the pest populations are under control is a strategy sometimes known as biological pest control. An example is the case study of California red scale described by Luck (1986). California red scale is an insect pest of grapefruit, lemons, and oranges in arid and semiarid regions. It inhabits all aboveground parts of citrus plants, inhibits fruit production, and kills branches. Red scale was introduced into southern California around 1870 on shipments of citrus nursery stock from Australia. Many potential predators of redscale were introduced from Australia. The most successful were ladybird beetles and parasitic wasps. Control was achieved in most coastal areas of California, but not in the San Joaquin valley.

Biological control using naturally occurring disease organisms can be used against weeds (McEvoy 1996). The common heliotrope which is a weed in Australia can be controlled through the introduction of a fungus (Hasan *et al.* 1992). Naturally occurring microbes such as root-colonizing bacteria can be useful in the control of plant disease, and some are currently available as a plant protection product (Kloepper 1996).

Ants, which show high diversity in traditional plantations, are also effective predators on insect pests (Carroll and Risch 1989). But ants also can interfere with insects that are beneficial to plants. A volatile chemical produced by young flowers of *Acacia* deters activity of ants, thereby facilitating visitations by pollinators (Willmer and Stone 1997).

Biological Control Using Natural History

Vampire bats range from tropical Mexico to northern Argentina and cost the Latin American livestock industry hundreds of millions of dollars per year. Early attempts at vampire bat control involved the use of traps, clubs, flame throwers, dynamite (for blasting their caves), the placement of strychnine at bite sites on cattle where bats were likely to bite again, and gassing of caves. The most effective control, however, has been based upon a knowledge of the natural history of vampires. Important characteristics of vampires are: they are much more susceptible than cattle to the action of anticoagulants; they roost extremely close to each other; they groom each other; their rate of reproduction is low; and they do not migrate. Injection of cattle with an anticoagulant that was then picked up by the bats resulted in a 95 percent reduction in areas where it was tested (Mitchell 1986).

Natural Insecticides

An important advantage of natural pesticides is that they are much less likely than synthetic pesticides to cause ill effects in vertebrates that become contaminated through eating treated plants and plant parts. Natural pesticides also do not seem to affect many pest predators (Stone 1992). Pyrethrum is a herbaceous plant in the composite family that has been used as an insecticide for 160 years (Bhat 1995). The flowers contain a mixture of pleasant-smelling esters that can be extracted and used to control insects. Compared to artificially synthesized insecticides, it has low toxicity to mammals and is rapidly degraded by ultraviolet radiation. Pyrethrum is one of the insecticides that may be applied to crops up to, and including, the day of harvest.

Because growing and harvesting the pyrethrum plant (*Chrysanthemum coccineum*) can be labor intensive (and thus expensive), it is not as economical as synthetic pesticides. However, chemists have not been able to artificially synthesize insecticides that are as environmentally benign. Consequently, pyrethrum still has a special niche. It is used as a broad-spectrum insecticide for use on minor crops, and as a quick "knockdown" spray on the day of harvest when high levels of chemicals are not acceptable (Silcox and Roth 1995).

The bacteria- and insect-killing properties of extracts from the seeds of neem (*Azadirachta indica*) have been known to scientists in India for almost a century. However, only recently have western scientists begun to explore the chemical nature of the seed extracts (Stone 1992). They have found that it wards off more than 200 species of insects, including locusts, gypsy moths, and cockroaches.

Ecological Management of the Crop Environment

Significant control of pests on managed croplands can be achieved by structuring the cropland to simulate the structure and species diversity of natural forests (Altieri 1987). For example, in traditional coffee plantations in Central America, coffee trees are grown under a canopy of shade trees. The structurally complex and floristically diverse traditional coffee plantations support a high density and diversity of predators and parasitoids. These are responsible for the reduced number of insect pests in traditional plantations compared to sun-grown monocultures of coffee. Web-building spiders also can be very effective predators on herbivorous insects. Coffee bushes in a traditional plantation had 34 percent more spiders than those in a monoculture (Perfecto et al. 1996).

Shading has been important in controlling the citrus rust mite in Florida. Groves planted in hammocks under oak and palm trees (which provide partial shade) have much lower mite populations than those grown in unshaded areas (Pedigo 1996). At high latitudes, increased shade reduces the hatching of spruce weevils. If trees are planted far apart, eggs in the open areas receive enough sunlight and high temperatures to hatch. When trees are planted close together, the increased shade reduces infestations (Pedigo 1996). A problem with close planting is that as the trees begin to mature, they crowd each other and become more susceptible to fire and disease. A better approach would be to interplant the spruce with a fast-growing "nurse" tree that would provide early shade. For example, scotch pine, a resistant species could be planted at 4-meter intervals, and after two years of growth, spruce could be interplanted between the pines.

Often, simply increasing the diversity and complexity of a cropping system helps reduce pests and disease (Liebman 1995). Beetles emigrated more from polycultures of corn, beans and squash than they did from monocultures (Risch 1981). Nematodes can be controlled by fallows or by alternating infested crops with other crop species. Orchards with a rich floral undergrowth showed a lower incidence of insect pests than clean cultivated orchards, mainly because of an increased abundance and efficiency of predators (Altieri and Schmidt 1985). In Kenya, intercropped maize and cowpea showed less damage from stemborers, as compared to monocultures (Skovgard and Päts 1997).

Other classes of fauna that are pests to crop species also can be influenced by diversity of crops or landscape. Deer, raccoons, and field mice — vertebrate herbivores that are troublesome for the farmer — can be controlled by coyotes and hawks in a diverse landscape.

Despite the fact that polycultures often have lower herbivore loads and a higher presence of predators upon herbivores, it is difficult to categorically state that polycultures always are more resistant to herbivores than are monocultures. One reason is that some herbivores are generalists, while others are specialists. If a pest insect is a generalist, increasing crop diversity might not decrease the impact of the pest. When some of the plants in polycultures serve as alternate hosts for insect pests, there could be little advantage over a monoculture (Andow 1983).

Another reason that evidence of polycultures being more pest-resistant is scarce is the difficulty of carrying out a definitive experiment. First, one must establish a statistically significant number of fields or plots that are exactly the same in all respects (such as soil type, soil depth, microclimate, previous history, topography, presence of soil organisms, and depth to water table). Then on some of the fields, one must plant a polyculture, and on other fields, monocultures of the species that make up the polyculture. Then, one must ensure that *all* plots are exposed to exactly the same amount of nutrients, water, and sunshine. Most importantly, all plots must be exposed equally to the same opportunity for insect infestation. This includes both harmful and harmless insects.

Logistically, this is very difficult. One must wait until the appropriate weather conditions occur. There must be a source of pests close by to allow the populations to spread. One might have to keep the experimental plots "ready" for years or decades before they became exposed to a potentially harmful population. It is little wonder, then, that definitive tests are rare.

It took just such a rare effort by a research team at Cedar Creek Ecology Station in Minnesota to demonstrate that the flora in high-diversity plots are less susceptible to grasshoppers than flora in low-diversity plots. The team planted 500 plots with varying numbers of 24 prairie species, in otherwise uniform plots in a prairie ecosystem. They had to monitor the plots for 11 years, but when an attack finally came, there was greater consumption of plants in low-diversity plots than in high-diversity plots (Culotta 1996).

Scientifically proving that diverse and complex ecosystems are more resistant to disease is even more difficult to prove. The dynamics of the vegetation and the insect vectors are complicated by another layer of factors that influence disease dynamics. Nevertheless, the Cedar Creek team that showed a diverse prairie ecosystem was more resistant to herbivores also showed that in low-diversity plots, aster plants were hit harder by a fungal pathogen than those in high-diversity plots (Culotta 1996).

Using Minimum Control

While controlling pests by working with nature is ideal, it is not always possible. Sometimes it may be necessary to rely on controlling nature. However, there are many degrees of control, and those that control pests with a minimum of impact on ecosystem function are preferable.

Monitoring

Monitoring cropping areas and spraying with pesticides only when conditions favor outbreaks of specific pests is a useful way to reduce the use of chemicals that have harmful side effects. One type of monitoring is that of weather conditions. It is useful when the combinations of heat and humidity that favor the outbreak of a particular pest are known.

Another type of monitoring is the use of traps baited with insect attractant, often sex pheromones. Crops are sprayed only upon evidence that pests are in the area. The Forest Service maintains a network of traps for the gypsy moth within its natural range, and chemical methods of control are used only when there is evidence of a potential outbreak.

In the state of New York, where onions are an important crop, the monitoring of onion thrips and spraying only when populations exceeded a critical level resulted in economic savings and environmental benefits (Hoffman et al. 1995).

Biological Control with Genetically Altered Organisms

Searching for naturally occurring agents of control or for naturally resistant strains of crop plants is often slow and sometimes unsuccessful. Genetic engineering may be more economically efficient. Bacteria, viruses, fungi and protozoa are being engineered so that they more effectively control insect pests through expanded ranges, more toxic mode of action, and faster kill rate.

However, genetic engineering poses risks (McEvoy 1996). The most serious risk associated with biologically engineered pest control is the lack of host specificity. There is always a danger that the engineered predator or parasite will attack species other than those for which it was intended. Misengineered bacteria could harm populations of honey bees. Fungi introduced to control weeds could adapt to crop plants. Insect resistance genetically engineered into plants could result in the elimination of beneficial insects (Federici and Maddox 1996).

Mass Trapping

Mass trapping relies upon the removal of sufficient adults to limit mating so that the population can no longer maintain itself. Mass trapping in Norway and Sweden apparently curtailed an outbreak of spruce bark beetles, and the death of pines central California was dramatically reduced following the trapping of the western pine beetle.

Mating Disruption

Mating disruption involves release into the atmosphere of large amounts of pheromone. Consequently, communication between the sexes is disrupted, mating is prevented, and populations are suppressed (Kadir and Barlow 1992).

Sterile Insect Technique

In the sterile insect technique, captive populations are grown in which the males are segregated and sterilized by radiation or other methods. When sterile males are released into the environment, they mate with females, resulting in unfertilized eggs. The approach is most effective with species in which the female mates only once. It has worked in the control of the screw worm, a serious pest of livestock in the

Southwestern United States and South America. Before substantial control of screw worm was initiated, it caused up to $100 million in losses each year (Pedigo 1996).

Photoactive Dyes

Light will catalyze toxic reactions in insects when dye molecules on or in their bodies absorb the photons of light. Recently, there has been a growing list of organic molecules which catalyze these same toxic reactions. At least two dozen insect species have been documented to be susceptible to photodynamic action of certain organic dye molecules. This area of research may now be maturing to the point where it can be used in practical insect control (Heitz 1995).

ECONOMIC ANALYSIS

Programs in weed control (Chapter Five) do not attempt to eradicate weed species from the earth, but merely try to keep levels of infestation at an acceptably low level. In contrast, the goal of some pest control programs is total elimination. Attempts to totally eliminate a pest can be successful only if every single mating pair of that species — anywhere in the world — is destroyed.

The economic threshold approach may be more rational. Here, the costs of killing a percentage of pests is compared with the economic damage done by the pests at the remaining population density. For every pest/crop combination, there is a break-even point, beyond which additional chemical application will not pay for itself in terms of the reduction in insect damage (Pedigo 1996). Just as in the case of weed control, however, it is very difficult to predict what that threshold will be. Further, long-term threshold levels may differ from short-term thresholds. Long-term thresholds will take into consideration the levels of pest populations that adequately support a predator species, which will be important in the event that pests arrive from other fields, or other regions.

ENVIRONMENTAL DISTURBANCE, AND PESTS AND DISEASE

It is difficult to find documented cases proving that human modifications of the environment have caused outbreaks of pests and disease in crop populations. It is easier to find examples of how human modifications of the structure and function of natural systems has provoked outbreaks of disease in *human* populations simply because cases where humans are impacted have been better studied.

Lyme Disease

In 1975, 51 cases of Lyme disease were reported from Old Lyme, Connecticut. Earlier cases were probably misdiagnosed as rheumatoid arthritis. Since then, Lyme disease has become the most common insect-borne disease in the United States. The Lyme disease pathogen is a spirochete, the same order of microorganisms that causes syphilis. It is transmitted by ticks from wild deer to humans (Real 1996).

The emergence of the disease in the 1970s is a consequence of landscape modification and disruption of natural functions that earlier controlled the disease. In precolonial and colonial times, wolves (and possibly cougars) were the natural agents that controlled the deer herd. As a result of forest clearing for agriculture in the early 1800s along the eastern seaboard of the United States, forest cover was greatly reduced. Deforestation, along with hunting, drastically reduced the deer herd, and the predators that preyed upon them were driven to extinction. However, during the late 1800s, agriculture shifted to the Midwest. Many eastern farmlands were abandoned and natural reforestation began. By 1980, the Northeast had four times as much second-growth forest than in 1860. Because there were no predators to control the deer herd, the increase in secondary forest stimulated a population growth of deer that exceeded the levels that existed before European colonization. At the same time, there began a much greater occupation of deer habitat by humans, due to suburban development and recreation. These factors facilitated the spread of Lyme disease into the human population.

To decrease the incidence of Lyme disease in humans, the deer herd should be reduced. Reintroduction of predators is one mechanism. However, predator control of deer in the east is probably not possible.

Although the red wolf is being experimentally introduced into some wilderness areas of North Carolina (DeBlieu 1992), wolves are incompatible with development in areas where Lyme disease is prevalent. Thinning of the deer herd through increased hunting is a feasible approach. Sterilization of deer through injection by dart guns is another.

Onchocercosis and Malaria

In 1909, when most of western Ecuador was covered with dense tropical forest, anthropologist S. E. Barrett wrote about how impressed he was with the lack of tropical diseases among the Chachi, a group of indigenous people residing in the tropical rain forest of northwestern Ecuador. He was particularly impressed because of the presence of the plague and yellow fever just two hundred miles to the south in Guayaquil (Barrett 1925).

Today, 97 percent of western Ecuador has been deforested (Gentry and Dodson 1992). Currently, the Chachi are severely infected with many tropical diseases, including onchocercosis, leishmaniasis, malaria, measles, tuberculosis, and typhoid (Medina 1992). There may be a cause and effect between tropical deforestation and incidence of these diseases.

Onchocercosis

Onchocercosis is a disease transmitted by the bite of female blackflies of the genus *Simulium*. The infective agent is a filarial worm. The adult worms are found in fibrous nodules of the skin, mainly around the head and shoulders. The female worm releases microfilaria that migrate through the skin toward the eyes where they can cause blindness (Benenson 1995). The blackflies prefer water in which there is sufficient organic debris to feed the filter-feeding larvae. Onchocercosis spreads following deforestation as a result of an increase in suspended organic matter in rivers and streams through the process of erosion. As soil erodes, it loses its ability to hold rainfall, leading to flooding — which, in heavy rainfalls, may join the river's courses and increase their turbidity. The Chachi, who customarily bathe in the rivers, become more exposed to the agent of transmission.

Malaria

Infective agents of malaria in Ecuador are species of the genus *Plasmodium* which are spread by the female *Anopheles* mosquito. *Anopheles* generally breed in standing water. Generation time is rapid, and high humidity increases the life span of the insect. The incidence of malaria is heightened by the increase in exposed standing water caused by deforestation (Desowitz 1987). Within undisturbed tropical rain forests, one rarely encounters stagnant water that facilitates the spread of malaria. Rain forest soils are highly porous and can soak up water before it accumulates. They are porous because the humus from decomposing wood and leaves helps bind the clay into particles that form a porous mass. When rain forests are cut down, the humus disappears, and the clay disperses into an undifferentiated and impermeable mass. Rain will then result in pools of stagnant water where the *Anopheles* can reproduce.

Because of deforestation, the local environment of the Chachi is now ideal for the quick generation and long-term support of the *Anopheles*. In addition, the Chachi live in open-sided houses near the river, which increases their exposure to the mosquitoes. As of 1994, the rate of malarial infection among the Chachi could be as high as 90 percent (Raich 1996).

The case studies of Lyme disease, onchocercosis, and malaria illustrate how the structure and function of natural ecosystems helps prevent the spread of pests and disease. When natural ecosystems are destroyed and replaced by cropland and pasture, the mechanisms that control the disease are also destroyed. The principle that natural systems help to contain the spread of pests and disease applies equally to pests and disease that affect humans as to those that affect crops and domesticated animals.

CONCLUSION

Crops, trees, and animals for commercial exploitation are bred and raised under conditions quite different than those that exist naturally. Under natural conditions, diseases usually do not decimate their hosts. But under economic cultivation, mechanisms of protection often do not exist, either because resistance has been inadvertently bred out, or the production system itself is more vulnerable to the spread of disease.

The question today for scientists and resource managers who are concerned about sustainability is how to reincorporate these functions and services of nature back into production systems — how do we go about rebuilding disease resistance into economic production ecosystems?

Integrated pest management in all its forms is necessary for the agriculture that supports modern society. Some techniques merely reduce dependence on overpowering Nature. A more sustainable way to increase insect and disease resistance in cropping systems is to incorporate the functions and services of nature back into the production system.

CHAPTER 8

INTERACTIONS DURING SUCCESSION

Succession is the replacement of one community of organisms by another. A series of successional communities (a sere) can begin on bare rock when a volcano rises from the ocean (primary succession) or on soil that has been denuded of vegetation (secondary succession). Secondary succession can begin after a natural disturbance, such as a landslide, or a human disturbance, such as clear-cut logging. Abandoned agricultural land is often the starting point for a successional sere.

FACTORS INFLUENCING SUCCESSION

Size of Disturbance

The plant communities that colonize a disturbed area and the sere that follows depend upon many factors (Jordan 1985). One is the size of the disturbance, because size influences how seeds are dispersed into the area to be colonized. In small disturbances, such as gaps in the canopy formed when a single large tree falls or is cut (a *treefall gap*), seeds from surrounding trees can fall directly into the opening. Consequently, the new community may resemble the surrounding community. In intermediate-sized disturbances of a few hectares, some seed dispersing animals freely traverse the disturbed area. For example, birds whose habitat is the closed forest might not hesitate to fly across a 1-hectare clearing, or perch for a few moments on a bush in the middle and defecate a seed. The community that becomes established depends partly on the food preferences of the animals that disperse the seeds. In large disturbances, usually greater than 10 hectares, the distance across the disturbed area is longer than the normal range of most animals which carry seeds. Revegetation in the middle of the disturbed area occurs through windborne seeds. The successional community will be different from the surrounding community if the latter consists of species that have heavy seeds.

Intensity of Disturbance

The successional sere also depends upon the intensity of the disturbance. A mild disturbance would be one that does not disturb the basic structure and function of the ecosystem. A treefall gap is usually a mild disturbance. The soil community and all of its nutrient recycling functions remain intact. Seedlings and saplings already growing in the opening usually survive. In fact, their growth may increase, due to the decreased competition from the older tree (or trees) that fell or were removed.

A moderate disturbance is one in which the above-ground structure of the forest is destroyed, but a functioning below-ground community remains. One example is when a large area of forest is blown down by a hurricane. Another is when a forest is cut down and immediately replaced with a tree plantation. In the latter case, the species composition of the below-ground community may change, but the flow of organic matter from litter into the soil is interrupted only for a short time, and nutrient recycling and conservation functions continue.

A severe disturbance is where both the forest structure is destroyed and the soil is severely degraded. Lava flow from volcanoes and forest clearing with heavy machinery destroy both the above-ground and below-ground ecosystems. Overgrazed pasture also can be a severe disturbance. The disturbance that gave rise to the successional sere shown in photoessay "Succession" can be considered severe. The field had been cultivated for grain or cotton for many decades, resulting in complete loss of the A horizon (topsoil). The remaining soil had very little organic matter and very few soil organisms that recycled nutrients and maintained soil structure.

Duration of Disturbance

The duration of a disturbance also is important in determining the nature of the community that becomes established. A single, discrete occurrence, such as a hurricane, may knock down the large trees in a forest, but may not injure smaller trees nor disturb the soil. Recovery occurs through growth of the seedlings present, or through the germination of seeds present in the soil before the disturbance. The replacement community may closely resemble the original one.

An intermediate disturbance would be one of a few years or less, during which time some — but not all — of the soil community is destroyed. Shifting cultivation, where the land is cultivated for two or three years and then abandoned, is an example. The replacement community would depend upon both the surrounding plant and animal communities, and the seeds and sprouts that remained in the soil when the plot was abandoned.

A long-term disturbance is one in which the disruptive activities continue for decades. Agriculture in the temperate zone is typically a long-term disruption. A farmer may till and cultivate his fields for years or decades. Recovery depends upon rebuilding the structure and function of the soil community. Each stage in soil rebuilding is accompanied by a discrete plant community (see photoessay "Succession").

TRENDS DURING SUCCESSION

The natural course of succession reestablishes the community and its functions that were destroyed by disturbance (Odum 1969). The length of time for recovery of a disturbed area depends upon the combination of size, intensity, and duration of disturbance, and the interaction between these environmental variables and the biota of the surrounding ecosystem. An opening in the forest caused by a small, light, short disturbance (such as the death of a canopy tree) will heal quickly, and within a few years, little evidence will remain of the disturbance. In contrast, a clearing of hundreds of hectares which is plowed intensely for many years requires a hundred years or more to rebuild a community that resembles the original.

The process of an orderly replacement of one community (including plants, animals, and soil organisms) with another is most apparent when the size of disturbance is large, the intensity severe, and the duration long. When succession begins in an area with a long history of intense disturbance, there is a distinct progression of community and ecosystem properties. There is an increase in diversity of all types of species. As more species of plants become established, complexity increases and more new niches become available for insects, birds, and other animals. With more species of plants, a wider variety of litter becomes available over a longer period of time. As a result, a more diversified below-ground ecosystem can develop. With a more diverse soil community, there is more opportunity for yet even more plant species to become established.

With the progress of succession, biomass of the plant and animal communities increases, resulting in greater production of leaves and litter. This, in turn, causes greater input of nutrients and carbon into the soil and stimulates the development of the soil community, which in turn improves the environment for growth of roots.

An Example of Succession

During the course of succession, communities change — from those small in structure, low in productivity, low in diversity and complexity, and inferior in sustainability, to those large in structure, high in productivity, high in diversity and complexity, and superior in sustainability. To understand how the services of nature facilitate these changes, we can examine an example of succession and the accompanying functional changes of a particular sere. The sere is illustrated in the photoessay "Succession."

The oak-hickory forest was a dominant community of the Southeastern Piedmont in pre-colonial times (White and White 1996). Indians burned the forest to keep the understory open, thereby facilitating hunting. The burning was usually light enough so that it did not destroy the soil organic matter in the topsoil. Sometimes this topsoil was a foot or more in depth.

During the nineteenth century, much of the Piedmont in the Southeastern United States was deforested and converted to agriculture, primarily cotton. It was the organic matter in the topsoils that constituted the "natural capital" that allowed development of cotton agriculture and the establishment of the Southern antebellum culture. But plowing and cultivation for cotton destroyed the natural capital. Soil fertility declined, and the boll weevil invaded. As a result of these factors, as well as the Great Depression, much of that land was abandoned from agriculture in the 1930s and 1940s. Following abandonment of agriculture on upland soils, a successional sere often followed. The sequence was common throughout the Southeast (Keever 1950), and even today, fields can be found in various stages of succession.

Immediately following abandonment of cultivation, annual or biennial grasses and herbs become established. The species that occur depends upon the time of year that abandonment occurred and the phenology of seed production of the parent plants. Weather is also a factor. Some of the most common first-year species are crabgrass, horseweed, and ragweed. The second year, asters and other members of the Compositae family become dominant. The third year, perennial grasses in the genus *Andropogon* (broomsedge) invade, and remain important for several years. The seeds of these early successional species are wind-dispersed.

The next stage is the invasion of the old fields by pine. Seedling establishment often appears to radiate from parent trees adjacent to the fields. Pines are predominant on the droughty, eroded upland soils, while hardwoods (such as sweet gum and tulip poplar) are often important on richer alluvial soils.

Low soil fertility may be a factor in the dominance of pine on the upland sites. These soils are extremely low in nitrogen, potassium, and available phosphorus, due to a century of agricultural exploitation. Pines have a competitive advantage in this situation because of their minimal need for nitrogen. Their needles last several years, and consequently, less nitrogen is required for the synthesis of leaves than is required by deciduous trees. Another advantage is their association with ectomycorrhizae that helps them acquire phosphorus. Phosphorus is held strongly by the iron in the mineral soil and is not readily available to the roots of most other species.

Pines can dominate old field succession from several years to several decades. However, pines regenerate very little under their own canopy. They are shade intolerant — that is, they need full sunlight to grow. Also, the seeds do not germinate well in the accumulated *duff* (coarse litter) on the forest floor. Another factor causing decline of the pine community is competition with hardwood species, such as oak and hickory. A layer of humus builds up on top of the soil under the stand of pine, and nutrients in this humus are more concentrated than they are in the mineral soil. This may be enough to give the hardwoods an advantage, especially with their more varied types of mycorrhizal associations.

Once the oaks and hickories become established, the quality of the soil begins to improve. An A horizon develops, in which the mineral soil can contain substantial (>2 percent) organic matter. With time, the horizon can be 6–12 inches deep or greater. Such horizons can be found in relic stands of old growth hardwoods in the Southeast.

Because oaks and hickories are relatively fire-resistant and readily sprout from the base when damaged, repeated burning results in arrested succession. If an oak hickory forest is protected from fire and if seed sources are available, shade-tolerant maples and beech will become established. They are better competitors under shaded canopy conditions.

Perhaps the most important service of nature during this successional sere is the improvement of the soil, particularly the increase in quantity and quality of soil organic matter. Initially, the soil is low in nutrients and soil organic matter, and has poor physical structure. The first herbs and grasses stabilize the soil and help control erosion. The pines lay down a substantial layer of organic matter on the previously bare soil, but it is low in nitrogen — thus not a particularly good resource for the community of soil organisms. As the hardwoods invade, the quality of the litter improves. Each community improves the environmental conditions of the soil, thereby facilitating an invasion by the succeeding community.

As the diversity of organisms increases, so does the complexity of ecosystem structure. The single-storied canopy of the pines is replaced by tall, spire-shaped sweet gums and tulip poplars, spreading oaks and hickories, an understory of dogwood and mulberry, and a shrub layer of buckeye and blueberry. Vines of wild grape, Virginia creeper, and poison ivy appear whenever an opening is formed by a blowdown. The more complex above-ground ecosystem is accompanied by a more complex below-ground ecosystem, resulting in more efficient systems of nutrient recycling and light utilization — yielding greater productivity (total ecosystem production of living matter) and greater sustainability.

The Nitrogen/Phosphorus Dilemma

Nitrogen is often a limiting factor for plant growth in early successional communities. Therefore, one would expect that leguminous trees with nitrogen-fixing symbiotic capability would be common in early successional seres, as the nitrogen-fixing ability would give it a competitive advantage. However, nitrogen-fixing trees are only occasionally dominant in tropical successional seres, and rarely occur in temperate succession. One reason may be that the ability to fix nitrogen is limited by the availability of phosphorus in the soil. The energy required to fix nitrogen subtracts from the total energy available to compete for other resources, such as phosphorus (Vitousek and White 1981).

The so-called "wonder trees" — popular with development agencies in the past for rehabilitating degraded lands in the tropics — are frequently nitrogen fixing species such as *Acacia mangium*, *Gliricidia sepium*, and *Leucaena leucocephala*. However, the nitrogen-fixing bacteria associated with these species need phosphorus, and the lack of this nutrient may be one of the reasons these species are sometimes susceptible to pests and disease when they are planted in poor soils.

FIGHTING SUCCESSION

As a result of the extensive abandonment of agricultural lands in the 1930s and 1940s, pine forests of harvestable age have been common in recent decades. Landowners usually sell this pine to pulp or lumber mills. Because these mills provide a ready market for the pine, landowners often wish to reestablish a pine forest. However, during the fifty years after pine became established in abandoned agricultural fields, soil conditions changed and no longer favor pine. The higher soil organic matter and its associated below-ground community instead favors the establishment of hardwood species.

Saplings of hardwood species usually already exist under the pine. When the pine is cut, they are "released" — that is, they begin to grow rapidly due to diminished competition. Mechanically cutting the hardwoods does little to suppress their competition with newly planted pine because the hardwoods resprout vigorously.

Landowners usually take one of two approaches to the problem (see photoessay "Forestry in the Southeast"). One is to use bulldozers and "roller-choppers" (large machines with blade-covered drums) to tear up roots of hardwoods, destroy the soil organic matter, and expose mineral soil. This essentially sets back succession in the soil to a point comparable to that when agriculture was abandoned. The other approach is to kill the hardwoods with herbicide. This approach preserves the soil organic matter, which is desirable for soil conservation. However, it does not destroy the soil organisms that tend to favor reestablishment of hardwoods.

The situation is a prime example of what is wrong with resource management — it tries to adapt natural systems to the constraints imposed by an artificial economic system. A more sustainable system of management would try to adapt the economic system to the constraints of natural systems. For example, instead of tearing up the soil to give pines an advantage, a more sustainable policy would develop markets for the hardwoods that establish naturally following pine.

The lack of understanding of the processes that occur in the soil during succession may have led to a misdiagnosis of the so-called "forest dieback" of coniferous stands in Europe. Dieback is frequently attributed to air pollution (Athari and Kramer 1983), and foresters sometimes experiment with additions of lime to the soil to combat the acidifying effects. While air pollution is certainly a serious problem, successional effects may also play a role. Many European conifers may be early or intermediate successional species. Changes may be occurring in the soil that favor later successional species, and these changes may be partially responsible for the observed dieback. The lack of replacement species in damaged conifer forests may be due to lack of seed sources for such species.

Foresters are not alone in their fight against succession. Almost all of agriculture is a fight against succession. Most agricultural grain crops are annuals — that is, early successional species. The farmer must continually fight against other successional species (weeds) that are better adapted to compete under disturbed conditions. Even most perennial crops, such as peaches and apples, are light demanders and do poorly in competition with later successional species.

The fight against succession is particularly difficult in the humid tropics, where conditions year-round favor the invasion of weeds. In temperate zones or in the dry tropics, cold or drought aids the farmer in his fight against succession by killing, injuring, or merely setting back the weed species. When the warm and/or rainy season begins, the weeds are just beginning their growth. Crops are not immediately overwhelmed. In the humid tropics, however, the growth of weeds never ceases. There is no "service of nature" such as cold or drought to set the weed population back to the point where it offers less competition to crops. That is why the establishment of annual grains and of pastures in the humid tropics is a particularly unsustainable activity. The fight against forces of nature — in this case, weeds — is especially severe.

Succession and Grazing Systems

Pastures are difficult to maintain where the natural mature vegetation is closed forest. Invasion of woody plants is continually a problem, because many woody species are better adapted at conserving nutrients in that environment than are grasses and herbs. Trees, with their deeper roots, are able to take up and recycle nutrients that occur sometimes up to 10 meters deep in the soil.

Annual burning of grasslands is a common method of controlling the advance of succession. While burning is effective in eliminating trees, it results in the rapid loss of nutrients. Calcium and potassium, which are bound in the organic matter, are changed into ash which is easily leached. Nitrogen is readily volatilized.

The problem of pasture degeneration is often worse in the humid tropics because of the rapid loss of nutrients in the continually hot and wet environment. In addition, woody species are able to invade all year round, and where there is no dry season, fire cannot be used as a control. Conversion of rain forest to pasture is the epitome of management struggling against nature. A tremendous effort must continually be dedicated to fighting against the natural successional forces that convert the pasture back to forest.

In general, grazing systems should occur in regions where grasslands are the naturally occurring mature vegetation. This is often in drier areas of prairie or savanna vegetation. In such areas, annual evapotranspiration is equal to or greater than precipitation, and there is little nutrient loss through leaching. Even in these situations, however, overgrazing can favor unpalatable shrubs. To avoid overgrazing, herds should be rotated from one area to another, in imitation of the dynamics of herds of wild grazers (Savory 1988).

There are exceptions to the principle that it is undesirable to graze livestock in areas where mature vegetation is forest. For example, areas of Kentucky and Florida are prime pastureland for horses, and the upper Midwest supports herds of dairy cows. Often these pastures are underlain by bedrock rich in calcium or magnesium, giving rise to soil in which pasture grasses are well adapted. The soils are rich despite the relatively high rainfall in these areas, and as a result, can support long-term grazing without serious loss of fertility.

WORKING WITH SUCCESSION

How can farmers work with, instead of against, succession? Following are some examples of techniques that minimize dependence upon artificial subsidies by taking advantage of facilitation during succession. They provide ideas upon which a more sustainable approach to agriculture can be developed. The techniques would have to be modified for different economic situations, but research on how to modify them and adapt them represents the challenge of sustainable resource management.

The Indigenous Model

The Kayapó Indians live in the Xingù River Basin in Brazil (Posey 1982). They begin a successional sere by clearing a circular-shaped field. Tree felling begins from the center and progresses outward. The fallen stems thus radiate outward like spokes of a wheel, and the bulk of the forest canopy biomass ends up near the perimeter of the circle. Corridors of relatively open areas lie between the tree stumps. Root crops such as yams, sweet potatoes, taro, and manioc are planted in the open corridors. The cultigens are already rooted and growing before burning occurs. Burning is carefully managed. A protracted burn minimizes the heat so that the root crops will lose only their green tops but not their viability. These pre-burn crops have a head start on weeds that will establish in the ash.

Papaya, bananas, cotton, and tobacco which require a high quantity of readily available nutrients are planted on the outer margins of the field, where ash concentrations are highest. A few weeks after the burn, men gather up unburned sticks and limbs, stack them in piles in various parts of the field, and set them on fire a second time. In the resulting piles of ash, other plants requiring high nutrients (such as beans, squash and melons) are planted.

These crops grow well for two or three years until soil organic matter is depleted and weeds become a problem. Meanwhile, rapidly growing trees — such as banana and plantain that were interplanted among the annuals — begin to close their canopy. Annuals and root crops are abandoned as the fast growing trees begin to bear fruit. These trees shade out weed species, and the leaves they shed begin to build up the soil litter that was lost during the burn, and during cultivation of annuals. In between the fast-growing trees, slower growing species that are useful for fruits, nuts, and medicines seed in naturally or are planted.

During this sequence, the productive capacity of the soil is scarcely reduced, even though fertilizers are not used. As the annuals begin to deplete the soil, the organic matter from the later successional shrubs and trees reestablishes the nutrient recycling and structural functions of the soil depleted by the annuals and root crops.

The Agroforestry Model

Indigenous agricultural practices are usually possible only where land is unlimited and labor has no cost. In some regions of tropical forest, land is not unlimited but is still plentiful. Labor is not free, but its cost is low. There, seres of economic crops that mimic natural succession are established by experienced farmers (Hart 1980). (See photoessay "Agroforestry in the Tropics.") They cut and burn the forest and plant maize and beans, as well as root crops like cassava. Interplanted with these are small trees such as cashew and citrus, and taller trees like coconut and rubber. Understory crops such as cacao are also established. Such a scheme is a good alternative for a farmer who owns a 100-hectare plot of ground. Every year or two, a clearing of one hectare is plenty for the needed supply of corn and beans. After a number of years, the farmer will have a whole series of plots, all in different stages of succession, and all producing different products.

Fallowing

In earlier times, when fertilizers were not as available as they are today, farmers would often *fallow* their fields — that is, abandon them to succession for a few years to partially reestablish their fertility. Fallowing also can refer to alternating a grain crop with a legume (such as clover or alfalfa) in a field in order to maintain soil quality. Because of the nitrogen-fixing capacity of many legumes, they restore soil fertility more rapidly than other species.

At higher latitudes where a winter crop is not possible, a grain crop may have to be alternated with a legume on an annual basis. In the south, however, some species such as crimson clover are moderately frost tolerant and can be planted in the fall.

Successional Mimics in Prairie Environments

Successional mimics culminating with trees usually are not economically feasible in developed countries because the income from trees is low compared to that from grains and vegetables. Trees are often relegated to land not suitable for agriculture. In areas where the mature community is prairie grasses, however, successional mimics may be feasible. Near Salina, Kansas, scientists at The Land Institute are experimenting with perennial polycultures that mimic the natural prairie ecosystem (Soule and Piper 1992).

This agroecosystem is modeled after the prairie. The seed crops are dominated by mixtures of perennial grasses, herbaceous legumes, and composites (sunflower and aster family). Individual communities would be composed of plants that differ in seasonal nutrient use and would complement each other in the uptake and release of nutrients. Each species has a distinct time interval in which it puts out leaves, then stems and flowers, and finally, fruits and seeds. As each species begins its period of domination, it appropriates large patches within the prairie ecosystem. Its leaves intercept most of the light, and its roots get most of the nutrients and water. But the period of domination is short, and by the next month, its dying leaves and stems are contributing nutrients and energy to the soil community that supports the subsequent flush of crop plants.

CONCLUSION

The problem in managing resource systems is that economically valuable resources often grow in early successional communities. Consequently, managers often must try to keep ecosystems in an early successional stage. However, early successional communities are not naturally sustainable. Without natural disruptions or chemical and mechanical inputs, natural tendencies such as increase in soil organic matter and increase in species diversity cause the system to evolve to a more mature state. Consequently, managing for short-term economic profit often requires a strategy that is different from that needed for sustainability.

For resource management to be sustainable, managers must look for ways to take advantage of facilitation — that is, the modification of the environment by one community of organisms to make it more advantageous for another community. The drawback to such a scheme is economic, in that a farmer or resource manager has to deal with a variety of crops. A farmer can be more economically competitive if he concentrates only on one crop.

There are other obstacles as well. An important cultural and economic characteristic of "developed agriculture" is the quest for structural and functional permanence in cultivated fields. Once an area has been set aside for a certain type of agriculture, there are strong cultural and economic forces that work to prevent change. If a farmer invests a lot of money in machinery to cultivate cotton, he naturally wants to keep his fields in cotton. Part of the resistance to change is also political. A farmer may be allotted a certain acreage for peanuts, based on past cropping history of the farm. This tendency of freezing into place certain cropping patterns discourages innovative cropping patterns based on successional patterns. In less developed regions of the world, agronomists and foresters are often freer to experiment with cropping based on ecological principles.

Our systems of agriculture, forestry, and range management will remain unsustainable as long as we insist on trying to force nature to accommodate the economic system. Only when we understand that it is more expensive to fight against nature than cooperate with her will we be able to achieve sustainability in our resource production systems.

PHOTOESSAYS

A. Succession: A Model for Sustainable Resource Management

Secondary succession begins when an agricultural field is abandoned from cultivation. The successional sequence shown here occurs on the Piedmont of Southeastern United States. Much of the agricultural land was in cotton during the 19th and early 20th centuries but was abandoned in the 1920s and 1930s due to the Depression and devastation of the cotton by the boll weevil. Since then, grain crops have sometimes been cultivated, but much of the land has changed back to forest. The soil is an Ultisol, meaning that it is very old, and highly weathered, with a low capacity of the clay minerals to supply and retain nutrients.

Figure 1. As a result of intense cultivation and the lack of a winter cover crop, the original topsoil (A horizon) has been entirely eroded away. At the beginning of succession, only subsoil is present. The upper 15 cm of subsoil (the "plow horizon") is tinged brown due to decomposition of roots from previous crops.

Figure 2. In the first year following abandonment from agriculture, the field is invaded by annual grasses and broad-leaved weeds. The annuals are replaced by perennial forbs (foreground and background) or perennial grasses, often *Andropogon virginicus* (center of photo). Fire favors grasses over the forbs and early successional trees, and periodic burning prevents the advance of succession.

Figure 3. In the absence of fire or other disturbance, pine generally invades the grassland. The pine often becomes established before hardwoods, because pine may be better able to tolerate the harsh microclimate and low fertility of the exposed subsoil. Seed source and dispersion are also critical in establishment of pine. The many stands of pine that today cover Georgia originated in this manner.

Figure 4. After 40–60 years, the pine stand becomes mature. Hardwood species such as oak and hickory become established beneath the pine and eventually grow up to replace it. A number of factors are involved: pine is shade intolerant and will not grow under a closed canopy; pine seeds do not germinate well in a thick layer of needles; and the type of mycorrhizae in the soil may change to a type that favors hardwoods instead of pine.

Figure 5. The mature oak–hickory stand of the Southern Piedmont is highly diverse and complex in structure and function. The oldest trees in this stand are about 150 years old. Leaves occur at all levels and angles within the forest and efficiently use the light entering the forest. The various species and size classes complement each other with regard to their need for nutrients and water. It has the highest net primary productivity of any naturally occurring terrestrial ecosystem in its range (when productivity includes wood, leaves, roots, fruits and nuts).

Figure 6. An important reason for the efficient functioning of the mature oak–hickory forest is the soil. The A horizon extends down to about 27 cm, and has a high content of soil organic matter that contributes to the supply of nutrients, as well as the ability of the soil to retain nutrients. Under natural conditions, it may take more than a hundred years for a deep, rich A horizon to develop. This rich A topsoil constituted the "natural capital" upon which the cotton industry of the 19th century South was built.

Figure 7. In the absence of fire, selective logging, or other minor disturbances, an oak and hickory stand will be replaced by beech (shown here) or maple, due to the greater competitive ability of beech and maple under low levels of light. The diversity and productivity of a beech or maple stand are less than that of the oak and hickory stand, due to the less complex — and therefore less efficient — structure and function of the beech (compare Figs. 5 and 7).

B. Forestry in the Southeast

Because of the prevalence of pine in the Southeast over the past half century, the market is well developed. In contrast, the market for hardwoods is much less developed because of the absence of supply. Therefore, when a landowner sells a stand of pine for pulp or lumber and begins reforestation, he generally replants pine. Another reason for choosing pine is its shorter rotation time (time from planting to harvest).

Figure 8. Site preparation for replanting of pine in Georgia. Note the understory of hardwoods in the background of this plot — a recently harvested pine stand being prepared for pine reforestation. The root rake on the bulldozer digs out the roots of the hardwoods, which otherwise will sprout and outcompete the pine. Sometimes herbicides are used to kill the hardwoods. It is necessary to fight against the natural tendencies of succession in order to get a second rotation of pine.

Figure 9. A conventional monoculture of pine, about 10 years old. The severe intraspecies competition for space and light is obvious. There is probably also severe belowground competition for nutrients and water. The stand needs to be thinned — an economically and ecologically inefficient operation.

Figure 10. A mixed-species plantation in the Piedmont designed for ecological and economic complimentarity. The light brown stems are *Paulownia fortunei*, the wood of which is valuable for furniture, moldings, and veneer. The rotation time is about 20 years. The white stems are *Populus tomentosa*, the wood of which is used for pulp. It is ready for harvest in about 6 years. The plantation is managed so that the poplar shades the sides of the paulownia, inhibiting branching of the paulownia and encouraging the vertical, cylindrical growth that makes trunks more valuable for timber. The paulownia and poplar will begin to compete at about 6 years, at which time the poplar can be cut for the market. The cutting acts as a thinning for the paulownia.

C. Alley Cropping in the Southeast

Alley cropping is a type of organic farming that is amenable to mechanization. The following figures depict an alley cropping experiment in the Georgia Piedmont. Details of the methods and data on nutrient cycling and productivity have been given in Matta Machado and Jordan (1995).

Figure 11. The alley cropping system in March. The leguminous hedge species, *Albizia julibrissin*, has not yet put out its leaves. The grain species in the "alleys" between the hedges is winter wheat.

Figure 12. By early May, the winter wheat has ripened and the albizia has put out its leaves.

Figure 13. After the winter wheat is harvested, grain sorghum is planted in the alleys with a *no-till* planter. Herbicides are not necessary, because the wheat has inhibited the establishment of weeds.

Figure 14. After the seeds have been planted, the branches of the albizia are lopped off. The leaves act as mulch, supplying nutrients to the soil and improving the microclimate. In this experiment, the branches were lopped off manually, but this could also be done mechanically by adapting a sidebar mower on a tractor.

Figure 15. By the time the sorghum is mature, the branches have regrown.

D. Sustainable Management of Forests in the Humid Tropics

Figure 16. The ideal model for sustainable production in the humid tropics is the naturally occurring rain forest. The diversity and complexity of structure and function are even greater than that in the mature temperate zone forest. Net primary productivity of tropical rain forests is greater than any other type of terrestrial ecosystem. This photo is from western Amazonia.

Figure 17. The high productivity of tropical rain forests is sustained by the efficient recycling of nutrients through a layer of decomposing organic matter on or near the soil surface. This layer is thickly impregnated with feeder roots from trees. Nutrients released from the decomposing matter are efficiently taken up by soil flora and fauna, or recycled into the roots. When the rain forest is cleared for conventional agriculture or pasture, the recycling mechanisms are destroyed.

Figure 18. The tropical rain forest can yield a sustainable harvest of economic resources when the products are extracted without destroying the structure and function of the natural forest. This photo is of a Brazil nut tree (*Bertholletia excelsa*), which gives an economically important yield.

Figure 19. This photo shows a Brazil nut collector slicing open the husk that contains the nuts. The biomass removed during a yearly harvest is a small percentage of the total forest biomass, but since total biomass is high, annual yield of a stand of Brazil nut trees can be considerable. Like topsoil, biomass is natural capital. With a large standing stock of biomass, absolute yield can be high even though "dividend rate" is low.

Figure 20. Palm fruits are an important extractive product of rain forests. This photo shows the fruit of the bacaba palm (*Oenocarpus bacaba*), used for oil and in making vitamin-rich drinks. Photo by C. Miller.

Figure 21. Animals, as well as plants can be extractive products. This is a black caiman, whose skin is valuable for leather products. Photo by E. Asanza.

There are a number of problems with extractivism in rain forests. The first problem is often developing a market for the products. Once a market is developed, the next problem is often the overharvesting of the product, leading to a depletion of the resource. Another problem is that when a market is developed, it becomes cheaper and easier to produce the product in plantations, thus driving the extractive harvester out of business.

Figure 22. In contrast to other extractive products, there is usually a good market for many tropical timbers. During harvest, trees are often cut and removed carelessly, destroying the ability of valuable species to regenerate. In some regions (as shown here in Venezuela), efforts are made at sustainable logging. Trees are felled in a direction that minimizes damage to other trees. Skid trails are carefully planned to prevent destruction of the litter cover on the soil. A precaution that should have been taken in the logging operation depicted is the cutting of vines several years prior to the operation so that they die and rot. Because vines in tropical forests often link together the canopies of adjacent trees, they pull down young trees when a large tree is cut for harvest.

E. Agroforestry in the Tropics

There are various types of agroforestry practiced in the tropics that increase the efficiency of land use, the efficiency of nutrient recycling, and the use of light.

Figure 23. *Taungya* is a type of agriculture that combines agriculture and forestry. Here, a peasant farmer plants a root cutting of teak in northern Thailand.

Figure 24. In between the cuttings of teak, the farmer plants upland rice. The weeding that he carries out (shown here) benefits both the rice and the teak. The intercropping of rice and teak can be carried out for about two years. By the third year, the teak begins to overtop the rice, and the farmer must then plant his crop of rice in a new plantation.

Figure 25. This photo of corn and sweet potatoes planted between rows of mahogany was taken in the Philippines. The leaves of the sweet potato cover the soil, prevent erosion, and intercept vertical rays of sunlight. The leaves of corn are better able to intercept horizontal rays. Orientation of the rows can be important. If the rows are east–west, the annual crops will get more sun in the early morning and late afternoon. During midday when the sun is overhead, the shade from the trees is at a minimum.

Figure 26. In an agroforestry plot near Tomé Açu in the eastern Amazon of Brazil, a cycle of cultivation starts with the clearing of secondary forest. Harvested trees are used for lumber and charcoal. After the remaining slash is cleared from the field, it is interplanted with corn seedlings and rubber trees. When the corn is half a meter tall, ginger is planted. Cotton, winged beans, and peanuts are planted in the same plot, following harvest of the corn. When the ginger is harvested, the husks — which the farmers claim repel nematodes — are spread on nearby beds of onions that are susceptible to nematodes.

Figure 27. Pepper is an important crop at Tomé Açu. It can be cultivated for 10 years before problems with rot force abandonment. It is interplanted with other valuable species, such as papaya (shown here), so that when pepper production declines, there is another crop already producing yield.

Figure 28. Multistoried plantations not only use light and nutrient resources efficiently, they are effective in protecting the soil. Shown here is an overstory of rubber trees and an understory of cacao. Latex from the rubber tree on the right is collected in a cup. A fruit of cacao is growing on the stem of the foremost tree on the left.

Figure 29. Often, a third component is added to the combination shown in Fig. 28. Here, a vanilla vine is shown climbing on the stem of a cacao tree.

Figure 30. The annual flooding of riverbanks by rivers carrying suspended soil is another function of nature that benefits sustainable agriculture. When the river rises, it floods and kills the weeds along the riverbank. When the river falls, it deposits silt that is rich in nutrients, and the farmer then plants his garden. The aboveground portions of the crops keep ahead of the weeds, while the roots are able to keep up with the falling water table.

Figure 31. The Mayans who inhabited portions of what is now Mexico and Belize practiced a type of agriculture now called *chinampas*. Canals are scooped out of swampy areas, and the mud is piled together to form islands in which crops are planted. Weeds growing in the resultant canals are collected and mixed with the mud to improve soil texture and drainage. Willows and poplars are then planted along the edge of the islands to stabilize the bank. This photo was taken near Mexico City, where flowers are cultivated in chinampas beds for national and international markets.

Figure 32. In the Philippines, a type of alley cropping is practiced in which the hedge species are planted along the contour. This arrangement has the additional benefit of reducing soil erosion. Shown here is an experiment in which erosion is compared in cultivated plots with and without contour hedges.

CHAPTER 9

THE TOP-DOWN APPROACH

AN ALTERNATIVE APPROACH

In the previous four chapters, we took the bottoms-up approach to designing sustainable systems. We looked at parts of the ecosystem — higher plants, herbivores, carnivores, insects, disease organisms, decomposers, and soils — and the interactions between these parts. We saw how these interactions provided sustainability in natural systems, and how these functions were the basis for economic production in management systems that do not rely heavily upon technology.

In this chapter, we take the top-down approach. We look at an ecosystem that appears to be sustainable and use it as a model on which to base a production system. We modify the existing system so as to take advantage of all the services of nature that it already incorporates, even though we may not know what all the functions and interactions might be.

Scientists argue about whether bottoms-up or top-down is better, when in actuality, both approaches are useful. The bottoms-up approach, however, makes up the bulk of modern biological and agronomic research. Most papers that are published in scientific journals describe mechanisms of how a system works, but very few examine or compare entire systems. Even in so-called "ecosystem ecology," where the focus is supposedly upon an entire system, almost all the reports concern a single process or function within an ecosystem. One reason lies in the reductionistic nature of the academic system. Another is the length of time and amount of resources necessary for a whole-system study, and the difficulty of designing a statistical test for a whole-system experiment.

Parameters for Whole-System Evaluation

There are several whole-system parameters for a top-down study of ecosystems. One is standing stock. This is the dry weight biomass of all living organisms in an ecosystem, including plants, animals, and bacteria in the soil. It is usually expressed on a per-hectare basis. Mature forest ecosystems may have standing stocks of several hundred tons per hectare.

Another whole-system parameter is species diversity. Diversity is the number of species that occur per unit area of an ecosystem. It is often expressed in terms of taxonomic or functional units, such as number of species of trees per hectare, or number of species of birds. Diversity has been examined as *within-habitat* diversity, *between-habitat* diversity, and *regional* or *global* diversity. Diversity within a farmer's field would be within-habitat diversity. Between-habitat diversity would be the diversity in a field plus in an adjacent patch of forest. Regional diversity would be the diversity in a biome, such as across the Alaskan tundra.

Other whole-system parameters concern ecosystem dynamics. Net primary productivity is the yearly biomass increase per hectare in woody plant parts plus the weight of leaves and fruit that are produced and shed. Net ecosystem productivity is net primary productivity plus biomass increment of animals per hectare. In successional ecosystems where there is a yearly increase in standing stock of plants and animals, net primary productivity and ecosystem productivity is greater than decomposition (the rate at which plant and animal remains on the ecosystem floor lose biomass). Decomposition can be expressed in tons of biomass per hectare per year, or in terms of a decomposition constant (Olson 1963). In mature ecosystems where standing stock of plants and animals change little from year to year, productivity and decomposition are approximately equal.

Nutrient recycling efficiency is the ratio between the rate at which nutrients are actually taken up by plants and the rate at which nutrients become available for uptake, either through natural inputs to the system or through fertilization. The efficiency ratio is in terms of kg of nutrients per hectare taken up by plants divided by kg of nutrients per hectare that become available through various natural or artificial processes. Highest efficiency occurs when biomass increment of the entire system is maximum, during middle to late succession. In mature communities, efficiency may decline (Vitousek and Reiners 1975).

Utility of Parameters

How can these parameters be used useful in designing a system for sustainability?

Standing Stock

In general, a system with a large biomass is more resistant to stress, and therefore more sustainable. An ecosystem with a large biomass is able to store nutrients, energy, and in some cases, water. Because of their large aboveground biomass, tropical rain forests are able to store calcium and potassium, despite the severe leaching potential of the climate (Jordan 1985). The implication for sustainable systems is that in a given climate and on a given soil type, a production system with a large biomass will be more sustainable than one with a small biomass.

Net Primary and Ecosystem Productivity

The first plants to invade an abandoned field are usually annuals. They devote a relatively large amount of the energy they capture to biomass and seed production, and relatively little to structures that ensure sustainability, such as woody biomass and a large root system.

Perennial herbs and grasses follow annuals during the course of succession. During their first year, they use much of their energy for production of woody and root tissue, and thus aboveground production may be less than by annuals. In subsequent years, the perennials draw upon stored energy, and their aboveground production increases.

Perennial herbs and grasses are replaced by trees, whose size is greater and architecture (spacing and distribution of leaves) is much more complex. Because trees are able to use more of the incoming light, they outcompete grasses and herbs and eventually replace them. Both aboveground production and sustainability increase during this phase of succession.

During late succession, the tree community may become dominated by species that dedicate most of their energy to sustainability functions such as resistance to disease, and relatively little to biomass production. Because economic production systems should be both productive and sustainable, the most desirable model for a production system is a mid-successional ecosystem.

Production of animal biomass varies during succession, depending upon the animal species, and the resistance of the plants to herbivory. Because of their high plant production, mid- to late successional ecosystems have the greatest potential to support large communities of animals. However, the potential may not be realized, because many plant species have evolved defense mechanisms (such as compounds that are toxic for herbivores).

Decomposition

When decomposition is greater than productivity in managed systems, sustainability is limited. An example of such a situation occurs when tropical forests are cut and used for annual crops or pasture. Since decomposition in these systems is greater than production, the stored capital (litter and soil organic matter) eventually disappears, and productive capacity declines.

Sometimes decomposition is lower than productivity, making the availability of nutrients a problem. For example, in old-growth spruce-fir forests of Canada and Alaska, needles accumulate in thick layers on the soil surface and immobilize nutrients necessary for growth of young trees.

In general, ecosystems should be managed so that the rate of nutrient release through decomposition approximately matches the need of plants to take up those nutrients. Synchrony of production and decomposition in natural systems occurs most frequently in mid-successional ecosystems.

Nutrient Cycling Efficiency

Nutrient cycling efficiency in annual ecosystems or early successional ecosystems is low because of the small root systems. As ecosystems increase in root biomass, recycling efficiency increases, and during mid-succession, nutrient recycling may be very efficient. Management for sustainability means managing for systems with well-developed root systems.

Species Diversity

Species diversity usually increases during the course of succession, maximizes during mid- to late succession, and declines somewhat during maturity. For example, in the photoessay "Succession," tree diversity is highest in the oak-hickory stage, and declines when succession proceeds to the beech "climax."

The higher species diversity in late successional ecosystems contributes to their relatively high productivity, because a diverse system uses resources such as light and nutrients more efficiently (Chapin *et al.* 1997). Diverse systems also are more stable than simple systems, when stability is defined as the ability to maintain a constant level of productivity despite climatic, biological, or human induced change or stress (Tilman and Downing 1994, Tilman *et al.* 1996). This is because change affects functionally similar species differently. A change in climate, for example, might increase the growth of one tree species and decrease that of another. When there is a redundancy of species in an ecosystem, the decrease in activity of one species can be compensated for by an increase in another species.

Optimal Model for Productivity and Sustainability

The best model for a managed ecosystem that optimizes long-term productivity and sustainability would not be an early successional community nor a climax community, but rather, a mid- to late successional ecosystem. Such a system would have high biomass, both above and below-ground. Although a proportion of the energy captured by such a system would be used for defense and maintenance, the economically valuable productivity would still be high because of the greater efficiency of the system. The system would have high nutrient recycling efficiency, and rates of decomposition would closely match rates of production. Finally, such a system would have a high diversity of species resulting in both redundancy and efficiency.

ENERGY AND POWER OUTPUT

An important fraction of the energy captured by some ecosystems is used for sustainability and maintenance functions and is not reflected in measures of biomass increase. Therefore, productivity is an inadequate measure of what an ecosystem is really doing. Power output is a better parameter because it includes functions of sustainability and maintenance, as well as those of biomass production. Power output is a whole-system parameter that integrates all of the whole-system measures into a single unit. It is the energy value of the biomass produced per unit time, plus the energy consumed in functions such as nutrient recycling. Power output of an ecosystem is equal to energy input minus heat losses per unit time. Power input to an ecosystem is determined through measurements of carbon assimilation by all its leaves throughout the growing season, and heat losses are determined by measuring respiration rates of all living organisms.

Carbon assimilation and respiration are difficult to measure accurately over an entire ecosystem for even a short period of time. Nevertheless, using power output as a theoretical measure of ecosystem performance can give us important insights into designs for sustainability and productivity.

H. T. Odum and Pinkerton (1955) hypothesized that maximum power output is the most important criterion for the survival of living systems. Survival of the fittest, they have said, means persistence of those forms which can command the greatest useful energy per unit time (power output).

Let us take a lion as a type of system (a lion's body being a system of interconnected subsystems such as the nervous system, arterial system, skeletal system, etc.). The lion through which energy flow is greatest will often win in the fight for a zebra carcass over another lion with less power output (a smaller and weaker lion). In other systems, maximum power output is a criterium of success as well. In a military system, that army in which the flow of energy is greatest (greatest amounts of food, ammunition, jet fuel, etc.) will be the army that, in most cases, wins the battle. Exceptions would occur in such cases as David and Goliath, where a large energy processing unit on one side is taken out early in the battle by a smaller, but more intelligent opponent (although H. T. Odum [1988] would say that intelligence has a very high value of embodied energy, and David was actually the greater energy processing unit). The hypothesis that the most successful system is the one that produces maximum power output can also be applied to ecosystems.

Power Output and Succession

Why does one successional community replace another? In chapter VIII, we discussed factors such as greater shade tolerance (more of the available sunlight can be used) and higher nutrient availability (due to functions of soil microorganisms). However, in the context of energy comparisons between ecosystems, late successional ecosystems replace early successional ecosystems because the former have greater power output.

The greater power output results, in part, from a greater biomass and a higher species diversity which result in a greater diversity in structure and function. This enables the systems to capture and use a greater proportion of the incoming solar energy. In small, simple systems, energy and nutrients are less efficiently captured and cycled, and as a consequence, simple systems have a lower power output.

Early successional ecosystems sometimes have greater rates of aboveground biomass accumulation than late successional ones, even though they capture less solar energy. This is because more of the energy they do capture is used for growth of leaves, stems, and fruit, and little is used for roots and for maintenance and defense. By contrast, not all the energy that is captured is transformed into biomass in mature systems. Some of it is used for nutrient recycling and other functions that ensure sustainability. Some of the energy is used for defense mechanisms, or to feed the tremendous variety of herbivores, predators, and decomposers as they go about processing and recycling energy, nutrients, and water (E. P. Odum 1969). While supporting these functions reduces the ecosystem's short-term productivity, it also increases the potential for future productivity. Thus eventually, the system with a greater power output will be a better competitor than one with a lesser power output, and the former will almost always replace the latter.

Ecosystem Evolution

An ecosystem is a group of interacting components. All species are connected — directly or indirectly — with all other species. The frequency of interactions between species determines the boundary of a particular ecosystem. For example, the interactions between species within a lake are more frequent than interactions between those species and species in a forest adjacent to the lake.

Ecosystems obviously change during the course of succession. They also change as a result of variations in the abiotic environment. Often, these changes are attributed merely to substitution of species. However, the idea that ecosystems maximize power output suggests that ecosystems also can change as a result of evolution — evolution in the same sense as evolution of species.

The idea that ecosystems evolve was a frequent topic of discussion among early twentieth-century ecologists. It was finally rejected by most scientists because ecosystems have no identifiable site (comparable to genes in individual organisms) where mutations could occur that result in evolutionary change. At the time that the idea of "ecosystem evolution" was discarded, very little was known or understood about mutualisms and other positively reinforcing interactions within ecosystems. Most evolutionary ecologists and population biologists focused on competition. In recent years, however, biologists have become much more aware of the numerous interactions that occur within ecosystems and how they arose through the process of co-evolution. Some modern ecologists now consider that all terrestrial higher plants, vertebrates, and arthropods are involved in one diffuse mutualism — and many are also involved in several (Janzen 1985). Evidence is also growing that interactions need not necessarily be direct. Often, indirect effects may be more important than direct effects (Patten 1991).

Species have a biotic environment: other species. If one species provides a favorable environment for another and vice versa, a positive reinforcement can lead to further adaptation and eventually cooperation. The ecosystem in which populations have evolved cooperative mechanisms will be more productive and more sustainable (have a greater power output) than an ecosystem in which each species continues to go about its own business, oblivious of other species except when it is necessary to compete for a limiting resource. Competition and selective pressure will favor ecosystems containing many mutualistic interactions, and eliminate those that have few or none.

For a theoretical example, let us take two ecosystems — one in which there is a mutualistic relationship between trees and mycorrhizal fungi, and another in which this mutualism does not exist. In the ecosystem with mycorrhizae, the trees can specialize in the capture of energy, letting the fungi specialize in scavenging for nutrients in the soil. When these two specialists work together, they can capture more energy and take up more nutrients than in the other system, where both the trees and the fungi have to be concerned with both energy capture and nutrient scavenging. Thus, where the two systems exist side by side, the ecosystem containing the mutualism will gradually outcompete and replace the other one that does not.

Lavelle (1996) provides a concrete example. He has shown how different types of decomposers in the soil interact with other organisms and decomposing soil organic matter to change the nature of the soil environment, and how this change results in another assemblage of species better adapted to the changed environment. He concludes that

> biological interactions in soils are highly dominated by mutualisms especially when abiotic parameters of the environment are not limiting. The establishment of such relationships may be considered as the result of a long period of coevolution which began with antagonistic relationships and gradually transformed into a global mutualism." (p. 218)

It is, of course, the species that evolve. But it is the simultaneous evolution of a network of species that results in the evolution of ecosystems. Species are not just evolving independently of each other. Each species makes up part of the environment of other species, and thus there is reciprocal influence. All species are evolving at the same time, under the influence of each other — and just as in classical evolutionary theory, competition is the force that determines success. It is the combination of species that captures the most energy and uses it most efficiently that is the most successful in the competition between ecosystems.

The fact that integrated systems of specialists are more efficient than collections of generalists is known from many different contexts. A team composed of interacting specialists — carpenters, plumbers, masons, electricians, and so forth — can build an urbanization of houses more efficiently than a group of handymen (who all have some knowledge about wood, pipes, wires and cement) who build an urbanization by each one building a house all by himself.

Gaia

As they adapt to their abiotic habitat, species influence and change that habitat, and the change in environment subsequently influences the species. For example, bare rock on a volcano may have been colonized by lichens, whose spores were carried in by the wind. As they respired, the lichens produce carbon dioxide, which dissolves in water to form carbonic acid. This acid breaks down the rock, forming soil. Once soil exists, the colonization by plants having roots becomes possible. As soils change, those plants best adapted to the new situation are the ones that survive and continue to change the environment. The idea that the flora and fauna of an ecosystem play an important role in the development of the soil is as old as soil science itself (Volobuev 1963).

The idea that species influence their environment and the new environment influences which new species will occur is very similar to the "Gaia" hypothesis, except on a smaller scale. The "Gaia" hypothesis refers to the claim that the biotic and abiotic elements of the earth form a coupled system which, by means of diverse feedback mechanisms, regulates the earth's atmosphere and temperature within limits conducive to the evolution of life (Rawles 1996). Gaia is also applicable at the ecosystem scale. Particular soils and bodies of water are not simply physical entities resulting from strictly geological processes to which species or organisms have adapted. Rather, the species themselves, through their metabolic activity, have influenced the physical and chemical conditions of the soils and water, and all have evolved together.

Gaia does not necessarily conflict with the basic tenets of Darwinian evolution (Margulis 1996). In Gaia, just as in classical Darwinism, species compete for resources. Some proliferate, and others become extinct. The important difference is that in the Gaia hypothesis, the species collectively are an important determinant of the environment. For example, about 2.3 billion years ago, oxygen was a great poison, and we would have expected its appearance on earth to be a major disaster for living organisms. Yet, as photosynthetic plants appeared and oxygen concentration in the atmosphere increased, consumer organisms evolved that depended upon oxygen for their metabolism, and thereby halted the further buildup of oxygen in the atmosphere (Lovelock 1996). The Gaia hypothesis also suggests that the biotic community plays a major role in keeping the chemical composition of the oceans and atmosphere from changing beyond the bounds within which life is possible, or changing faster than the rate at which species can adapt to the changes.

Caveats

To claim that ecosystems evolve, it is not necessary to claim, as Lovelock does, that the earth is a living organism. It is enough to point out that the earth and the forests, prairies, and lakes that inhabit it are not merely repositories for random assortments of species, but are integrally connected with the species that inhabit them. The interacting species and their substrate are systems in the most technical sense of the word.

Another important caveat is that a particular natural ecosystem may not be "optimally" adapted to its physical environment. There may be another combination of species which, if it occurred in the same physical environment, would be better adapted and would eventually replace the first.

A third important aspect of ecosystem evolution is that co-evolution of species does not necessarily have to take place in the ecosystem in question. For example, if we observe that squirrels are important in the regeneration of stands of white oak trees, it does not necessarily mean that squirrels co-evolved with white oak. The evolutionary history of squirrels may have been more closely bound with chestnuts. As chestnuts disappeared, the squirrels found their way into stands of less tasty, but nevertheless adequate white oaks with their acorns. Only when the squirrels happened to wander into a white oak stand did the mutualistic interaction between squirrels and white oaks begin.

Power Output, Efficiency and Managed Systems

Natural systems evolve toward maximum energy capture and efficient energy use, and thereby maximize their power output. On the basis of physical arguments, H. T. Odum and Pinkerton (1955) predicted that maximum power output of a system will occur when 50 percent of the energy captured by plants is used for increasing biomass and 50 percent is used for maintenance. They argued that in the long term, this type of system will outcompete other systems with different ratios of productivity to maintenance. Therefore, the ideal model for managed systems is one in which energy use is divided approximately half for production and half for maintenance.

Energy Analysis in Human Dominated Systems

While power output is useful in understanding natural ecosystems, it gives little guidance to designing human-managed systems where a portion of the power input is energy in the form of fossil fuels or products derived from the use of fossil fuels. In contrast, energy analysis is a practical guide to designing human-managed systems that minimize environmental problems.

Grain, wood, and beef (or any other output of farms, forests or pastures) are the product of three sources of energy:

1. the sun through photosynthesis;

2. petroleum energy as manifested in mechanical, chemical, and genetic technology;

3. human labor and intelligence derived from both 1 and 2.

Naturally occurring ecosystems derive all their energy from 1. Managed systems derive part of their energy from 1, and are subsidized by 2 and 3. It is subsidy number 2 that has been decreasing sustainability and causing pollution. Ratios of energy output to energy input for various systems reflect the degree to which the system is subsidized by fossil-fuel energy.

Paddy rice systems in Asia produce 20 calories of grain from 1 calorie of labor subsidy (Steinhart and Steinhart 1974). The ratio is high because of the high input from services of nature, such as nitrogen fixation by algae floating in the paddy and a fluctuating water table. Swidden agriculture, similar to that practiced by the Kayapó, yields approximately 12–15 calories of food from 1 calorie of energy input, mostly labor (Pimentel and Pimentel 1996). Efficiency is high because of the reliance upon soil enrichment through litter from trees. Traditional agriculture, as practiced on small family farms, results in about 10 calories of food (corn, potatoes, or beans) from one calorie of manual and animal labor. Efficiency is relatively high due to reliance upon services such as recycling of nutrients in cattle manure.

For high-intensity production of grain crops dependent upon energy subsidies such as fertilizers and pesticides, the output input ratio is only 2.5/1. While mechanical and chemical increase the efficiency of human labor, they decrease the efficiency of total energy use. It is the inefficiency of energy use that causes pollution and reduces sustainability. Energy-efficient systems rely heavily on the services of nature. Such systems resemble the structure and function of natural systems, and require less of the artificial energy subsidies that cause environmental problems.

GUIDING PRINCIPLE FOR RESOURCE MANAGEMENT

To optimize the productivity and sustainability of managed systems, we should design them systems according to the following top-down rule: **Farms, forests, cattle ranges, and other resource systems should be managed, as much as possible, to resemble the structure and function of the ecosystems or successional seres of ecosystems that occur naturally in the region.**

The more a managed system resembles a natural system, the more the system will rely upon the services of nature. The more that production systems rely upon the services of nature, the less they have to rely on mechanical and chemical subsidies that are the root cause of all the environmental problems in agriculture, forestry, and range management.

Application of Principle

A good example of a commercial production system that resembles a natural ecosystem of the region (and thus works with nature) is a pecan orchard in Georgia. Pecan orchards have many structural and functional similarities to the oak-hickory forest, a common natural ecosystem in Georgia. The pecan trees perform the function of nutrient storage and recycling. The orchard needs to be fertilized only with the amount of nutrients lost through harvest. Consequently, it requires a relatively low energy subsidy.

An example of a commercial production system that does *not* resemble the natural ecosystem (and thus fights nature) is a pasture in a tropical rain forest. The Amazonian pasture has to be heavily fertilized after the first few years following the cutting of the native forest, because once the forest is replaced by grass, the nutrient stocks are quickly leached away. If fertilization is not practical, invasion by woody shrubs and trees poses a severe problem, because trees are more effective at storing nutrients and taking them up when they have leached deep into the soil (Jordan 1985). It is a continuing struggle to keep a pasture from being re-invaded by trees; pastures in areas of natural forest always result in a fight against nature.

Grazing systems are best established in areas of natural grassland, because the food chain from grass to cattle resembles the structure of the food chain from grass to wild herbivore. Natural grasslands often occur where annual rainfall is low or concentrated in one season. Pastures in areas of natural prairie can be maintained more easily. Often, fire helps maintain the vigor of grasses by burning off the dead leaves and allowing sunlight to penetrate to the buds at the base of the leaves. Many grasses native to prairies have extensive underground root systems that supply organic matter to the soil ecosystem.

In lesser-developed regions of the world, there might still be opportunities to guide resource management along a path more in harmony with nature. Unfortunately in the "developed" countries of the world, it is now not possible to look at the natural ecosystem and then decide which crop would be optimally suited for that situation. In most cases, an economic system has already dictated which crops are planted, or which management system is used. The best that we can do is to take the cropping system that already exists and try to modify it so that its functions are carried out more by natural processes rather than by mechanical or chemical subsidies. Naturally occurring ecosystems can give us clues as to how to modify production systems, so as to take advantage of the services of nature.

CHAPTER 10

CASE STUDIES OF RESOURCE MANAGEMENT

Peoples in "undeveloped" economies have little control over their environment. They live under subsistence conditions, but subsistence does not necessarily mean unsustainable. In fact, cultures of some indigenous peoples in subsistence economies have survived hundreds and sometimes thousands of years. But since they could not control nature, they have had to work with nature in order to survive. They had to adapt to the natural systems in which they found themselves.

In general, "development" means a transition from a subsistence economy to a market economy where the belief is that humans have conquered nature. Adapting to nature is no longer thought to be necessary. In some cases, the transition to a market economy occurred through the elimination of the indigenous populations in their native habitat. In other cases, the local populations were enticed or forced to change their methods of resource production and extraction. In the first part of this chapter, we give several examples of the transition from subsistence management systems that were sustainable to market-driven systems that were not.

Despite the pressure in market-driven systems for short-term exploitation, there have been scattered attempts throughout the world to reverse the trend toward unsustainability. In fact, there have been cases where managers or local communities have realized the negative effects of unsustainable practices, and have developed and implemented more sustainable resource management. Examples of this transition are given in the second part of this chapter.

TRANSITION FROM SUSTAINABILITY TO UNSUSTAINABILITY

Agriculture

The Sahel, Africa

The climate in the Sahel region of Africa — in what is now Mali and Niger — is very dry. Average rainfall is less than 600 mm per year, a lower limit for most agricultural crops. As a result, the welfare of the Marka, a local ethnic group who are experts in the cultivation of rice, is highly influenced by climatic fluctuations. They have been cultivating native rice since prehistoric times, and they make complex and sophisticated decisions about when to plant and what varieties to plant (McIntosh 1993). Their decisions are influenced by environmental clues — different varieties of rice have different vegetative periods, different adaptations to various flood depths, flood timing, pH tolerance, and fish predation. Different varieties are sown at different time intervals on different soil types.

The knowledge that the Marka possess about rice and its cultivation is secret, and has been developed over a long period of time. It is a means of maintaining a specific ethnic identity. Social relations with other groups have become instituted as "buffering mechanisms" against potential bad times, allowing trade to occur without the necessity of immediate equal compensation. This buffering is useful, for example, with the Bozo fishers, who reciprocate labor, goods, and services with the Marka. The buffering is beneficial to both groups, because weather that favors one group may disfavor another.

Another important aspect of the Marka system of sustainable rice production is prioritized tenure on property held in common with the entire ethnic group. A hierarchical system prioritizes access to land, and the rules regulating access to common property has been encoded in Islamic law. Prioritized access ensures that those with the specialized knowledge are those that make decisions on varieties of rice to be planted, as well as the timing of the planting (Park 1992).

Ensuring sustainability of rice production requires a deep understanding of how various social systems work. Social systems have deep ties to the environment through culturally mediated and specialized relationships (Halstead and O'Shea 1989). To know the physical needs of a particular crop is not enough information to produce consistent quantities in a sustainable manner. Farmers make decisions based on variables which may seem "unscientific," because the farmers are considering these variables from a different temporal and spatial scale than normally understood in the developed world. One needs to understand the evolutionary nature of "secret knowledge" and intergroup relations which function together as part of a subsistence system, and which buffer the system against environmental and political variability.

Regional development projects in the Sahel break down this system of rice production. The goals of regional development projects are to increase the national market economy through production of rice, and to technically mediate the uncertainty of the climate so as to guarantee a more steady supply to the market (Ba and Crousse 1985). Development programs usually require cultivators to change from indigenous rice varieties to the Asian variety *Oryza sativa*, which is more marketable and has a higher yield, but requires consistent amounts of water. The knowledge about Asian rice is held by outsiders and is not secret. As a result of development, prioritized tenure on commonly held lands is eliminated, and equal access is gained by those without the "secret" environmental and social knowledge. The traditional allocation system, which is built on the recognition of natural variables, is replaced by a system organized to suit the demands of a capitalist economy.

Development goals of doing away with uncertainty and increasing production result in ecological deterioration. For example, in Senegal, 40,000 hectares put under irrigation for rice are now degraded, as inexperienced people quickly erected poorly built irrigation structures in order to satisfy government requirements for establishing tenure. Polders constructed to control water flow are not flexible enough in times of drought. Polders also affect fishing, as changes in the flow of the river and the displacement of water through polders affects fish breeding and feeding. The transition to a market economy ignores the nature of the Sahelian climate and soils, and deprives traditional Marka groups of their ability to respond flexibly in times of distress.

As a result of development, both the Marka and the Bozo have formed relationships with the state government and the development agencies. Traditional knowledge of cultivation and fishing are being lost. Myths may be changed, or no longer told at all, as individuals are drawn into the market economy and group identity becomes less important. Rice production increases in the short term due to artificial subsidies required by production of white rice. Long-term sustainability of rice production is lost as native species and the knowledge of their cultivation is abandoned.

Traditional groups such as the Marka and the Bozo worked with each other and with nature to ensure the sustainability of their culture and of their livelihood. Development required that the native groups abandon cultivation systems that worked *with* nature and take up systems with artificial requirements to fit an economic system. While this resulted in great profits for groups well integrated in the national and international economy, it has done little for the betterment of the local people and for the long-term sustainability of agriculture in the region.

Tropical Forestry

To gain an encompassing view of the transition from sustainable to unsustainable forestry in the tropics, we present here several examples of forest management along a gradient from sustainable to unsustainable.

Extractive Systems

The most sustainable management system is one that has little or no effect on the structure and function of the native forest, or that is itself a minor modification of the original existing natural system. Extractive systems where natural products such as gums, latexes, syrups, nuts, vines, and oils are harvested are examples (Dubois 1996). Animals also can be sustainably extracted from tropical forests. Parrots and fish are used for the pet trade, and reptiles are harvested for their skins (see photoessay "Sustainable Management of Forests in the Humid Tropics").

The key to sustainable management of extractive systems is in limiting harvest to a rate not greater than that at which the ecosystem can replace the harvested products. Here is where the system of extraction often fails. For an extractive system to survive, there must be a market for the products. Once a market develops, there is an incentive for more producers or harvesters to enter the field. As the market infrastructure develops, more people become dependent upon the harvest of the resource, inevitably leading to over-exploitation. For example, wild-harvesting of parrots in the new world tropics is being carried out at a rate greater than that of replenishment. As a result, Beissinger and Bucher (1992) have recommended a moratorium on importation of parrots from all wild-caught birds into the United States. The irony of extractive systems is that the seeds of destruction are an inevitable part of its success.

Another problem with extractive systems is their "economic inefficiency." In the name of conservation and sustainable resource use, areas have been set aside as "extractive reserves," and commercial harvesting is limited to non-timber products. In recent years, extractive reserves have been set aside for rubber tappers in Brazil (Browder 1992). However, harvest of rubber in these "seringals" is very labor-intensive. Rubber trees are dispersed throughout the forest, and rubber tappers must spend a lot of time walking between trees. It is difficult for them to compete against the monocultural plantations in southeastern Asia, which require much less labor. Brazilian seringals can survive only if they are subsidized by the government, or if the workers are content to receive a very low wage. Other rain forest products such as chocolate, quinine, guarana and allspice once were harvested from the native forest, but competition from plantations also forced extractive collectors out of the market (Nations 1993).

Selection Cutting

In selection cutting, whole trees are removed from the forest. When it is done carefully (and only a small percentage of all the trees are cut), the basic structure and function of the forest remains intact, and no long-term damage is inflicted upon the system. Processes disturbed by the harvest can recover quickly and easily, aided by spores, seeds, and animals from the surrounding intact forest.

During the early part of the twentieth century, selection cutting management systems for naturally occurring tropical forests were developed. In Africa, the system was often referred to as the "tropical shelterwood system" (Fox 1976). It was designed to promote the establishment, survival, and growth of the seedlings and saplings of desirable species by poisoning undesirable trees and removing vines and weeds. After several years, when surveys showed that the reproduction of desirable species was well established, the canopy trees of the timber species were harvested. The tract was then monitored, and weedings and thinnings took place as needed. The system was abandoned in the 1960s, partly because it did not make sufficiently intensive use of the land to compete with other forms of land use such as cocoa, oil palm, or agricultural crops (Lowe 1977).

Strip-Cutting

One problem with selective harvest is logistical. It is difficult and expensive to open up logging trails through the forest to harvest single trees, and even more difficult and expensive if care must be taken not to injure the remaining trees. Strip-cutting is a compromise that combines the ecological advantages of small openings in the forest with the logistical advantages of clear-cutting. Strip-cutting resembles an elongated tree-fall gap. A trial project in forest management utilizing strip-cutting was started in the early 1980s in the Palacazú Valley at the eastern base of the Peruvian Andes, and was supported by the U.S. Agency for International Development.

Approximately 3500 Amuesha (Yanesha) Indians inhabited the valley, where they practiced traditional shifting cultivation of manioc, maize, and upland rice. Hartshorn (1990) estimated that the surrounding forest contained 1000 species of native trees. Timber exploitation under the plan was carried out in long narrow clear-cuts, interspersed in the natural forest. Each strip clear-cut was 30 to 40 meters wide, and the length was determined by topography and logistics. Each strip was an elongated gap, bordered on each side by intact forest that supplied seeds for natural regeneration within each clear-cut strip. The plan was that in successive years, new strips would be located far enough from recently cut strips, so as to ensure adequate stocks for reproductive trees to repopulate the harvested strips. Regeneration also occurred through the sprouting of any remaining stumps.

The strip-cutting system was well-adapted to the local environment, both culturally and ecologically. Oxen or water buffaloes were used to extract the logs, poles, posts and fuel-wood. Extraction with draft animals is cheaper, and the logging and skidding damage to the soil is much less than with traction-driven tractors or skidders. Exceptionally large logs were sawn lengthwise in the forest with special portable saws.

The project did not survive very long after the U.S. aid came to an end. Part of the problem was caused by terrorist activity along the road leading to Lima, which made it difficult and dangerous to deliver output to markets. Another problem was that the price of timber was artificially depressed by government policies (Southgate and Elgegren 1995). These authors concluded that in order to receive the full market value for their timber, they would require substantial help with marketing and processing operations.

Plantation Forestry: Jari

In 1967, one of the largest conversions of tropical forest to pulp plantation began near the junction of the Jari and Amazon rivers in the state of Pará, Brazil (Jordan and Russell 1989). The project was initiated and financed by Daniel K. Ludwig, one of the world's wealthiest men and owner of numerous international corporations. Ludwig had anticipated a global shortage of wood fiber for pulp, and to meet this shortage, he and his advisors began looking for a site that would have high potential for pulp production. The most important reason for selecting a site in the Amazon was Ludwig's belief that the Amazon region had a great and unlimited potential for high and sustained yield of commercial crops.

Ludwig wanted not only to raise trees for pulp, but also to develop an entire integrated pulp production project that included a mill suitable for producing high-quality pulp ready for delivery to overseas markets. The project included a port on the Jari river, a railroad system, and hundreds of miles of roads. A complete town called Monte Dourado was built, including schools, medical care units, a supermarket, bank, churches, houses, dormitories, central cafeteria, motor pools, machine shops and a fuel depot. A series of outlying "silvavilas" was built to house and feed workers in remote areas of the plantation. Air service, complete with airport and planes, was established to ease the problem of slow access to and from Belem, 350 kilometers to the East and the closest port of access. The single most expensive item was the pulp mill itself, built in Japan for $400 million and floated across two oceans and up the Amazon river. By 1981, the total investment was approximately $1 billion.

Ludwig's advisors recommended melina (*Gmelina arborea*) as the best species to plant at Jari. Ludwig ordered the entire area planted with melina without ever having his advisors examine the soils at Jari for their suitability for this species. Only after growth of melina was 40 percent below target were the soils examined. In half of the plantation, soils were found to be too infertile for melina. As a result, trials with pine and eucalyptus were begun. Studies of productivity and nutrient cycling at Jari (Russell 1983) indicated that stand productivity at Jari was relatively low compared to other plantations of the tropics, and lack of soil fertility was an important contributing factor. Other factors appeared to be inappropriate machinery, inadequate worker training, poor site preparation, inferior genetic stock, weeds, pests, diseases, and fire.

In an International Workshop of Forest Management in the Tropics, Palmer (1986) presented an independent analysis of the factors which caused the low profitability at Jari. He included:

1. Loss of surface soil and compaction of subsoil due to site clearing with heavy machinery.

2. Planting of melina on infertile soils, where it is incapable of good growth.

3. Machine removal at $500/ha of failed melina stands.

4. Less than optimal spacing of trees.

5. Delay in trials with pine and eucalyptus.

6. Failure to utilize native species to the extent possible.

7. Necessity for expensive fertilization.

8. High incidence of pests and diseases.

9. Fires set by disgruntled employees.

10. Social and cultural factors such as a wholly owned foreign company in a relatively remote part of Amazonia; a company with headquarters at an unfortunately named "Gold Hill" in a country fascinated by gold; rigorous controls by guards on access to the estate.

These features aroused suspicion and dislike in increasingly nationalistic Brazil, seeking to distance itself from new forms of colonialism. The prolonged reluctance of Ludwig to engage in dialogue with even the more responsible members of the press was apparently due to his belief in the power of his personal contacts with an increasingly tottery military regime.

In 1982, a majority interest of Jari was sold to a consortium of 27 Brazilian companies for $280 million. The question remains as to who finally paid for the $720 million loss sustained by Ludwig when he sold Jari to the Brazilian investors. Ludwig's parent corporation, International Bulk Carriers of New York, may have been able to take the loss as a tax deduction against U.S. income tax. If this was the case, the U.S. taxpayers paid for part of the losses at Jari. If Ludwig defaulted on any of his debts to Brazilian banks, then the Brazilians themselves paid in part for the destruction of their own native forests.

Jari has not contributed to social stability in the region. In 1987, turnover of chainsaw operators averaged 90 days. These workers leave their families in other regions of Brazil, take a job long enough to make some money, and then return home. Turnover of technicians and engineers also was high, with 50% leaving every year.

In 1983, the Jari reported a $2.1 million (U.S.) operating profit (International Society of Tropical Forestry 1985). This represents a rate of return of 0.2 percent on the original investment of 1 billion. This rate of return is insufficient to draw further private investment, and thus the project is economically unsustainable in the marketplace. It continues through the benefits of government subsidies (McNabb *et al.* 1994).

Exploitative Logging

The most common form of timber harvest in the tropics cannot be called forestry at all, but rather exploitative logging. In some countries (Brazil for example), the process has often begun with the construction of roads, financed by national governments or international banks. The purpose of the roads is to open up a forested area for colonization, other types of development, or to ensure territorial sovereignty. Logging quickly follows, because the roads provide easy access to timber. In other countries (Malaysia for example), "logging concessions" are granted to large companies by national governments. In either case, the most common type of harvest is "high grading" or "creaming," in which only a few of the most valuable trees are taken, and the rest of the forest is left degraded due to careless logging practices. Often, only five percent of the trees are taken, but one third to two thirds of the remaining trees are damaged beyond recovery. Commonly, almost a third of the ground may be left bare, with the soil impacted by heavy machinery (Myers 1984). Once a forest has been opened up by such damaging activities, fires could easily destroy the rest (Uhl and Buschbacher 1985).

Range Management

The Serengeti

The Serengeti ecosystem straddles the border between Kenya and Tanzania in East Africa. It contains approximately 30 species of ungulates, 400 species of birds, and 13 species of large carnivores, with numerous other smaller forms. The wildebeest is the most numerous large mammal. They perform characteristic seasonal movements called the "migration" — from the short-grass plains (where they calve in the wet season) to the western corridor in the early dry season, and finally to the north for the late dry season.

The characteristic migration is in response to the extreme temporal and spatial patchiness of green forage. Green areas are commonly widely separated in space at any given time and are highly unpredictable. Unpredictability of green forage is caused by the unpredictability of the climate. Because of the nomadic behavior of the wildebeest, they are adapted to exploit the widely separated bursts of grassland productivity wherever they occur, and their dense herding behavior allows them to crop the grasslands in a way that increases forage yield.

The stimulation of native grasses by grazing is a consequence of co-evolution of grazers and Serengeti grasses. An important characteristic of these grasses is the ability to maintain a high enough shoot biomass to maintain root reserves. Cultivated grasses would not be able to sustain their growth under the severe defoliation regimes characteristic of this grazing system. It is the mobility of the major grazing animals, coupled with the ability of the grasses to sustain themselves under periodically intense grazing, that accounts for the ability of the Serengeti to support an unparalleled grazing fauna (McNaughton 1979).

There are complicated interactions between the forage species and the herbivores, between the various species of herbivores, and between the herbivores and predators. Wildebeest affect vegetation by altering the competitive balance among plants. Zebras are able to digest low-quality forage, and are less affected than wildebeest by the changing palatability of grasses. Grant's gazelles eat herbs almost exclusively and are tolerant of dry conditions, enabling them to remain on the plains for a large part of the year. The giraffes prefer to eat bushes and trees. The wildebeest is the main food of the hyena and lion. However, these predators are confined to fixed territories, so when the migrants leave, the predators have to switch to less abundant resident prey. The cheetah population on the plains increased, possibly in response to increasing numbers of gazelles, which have responded to the increased presence of herbs resulting from increased precipitation in the early 1970s.

There are continual fluctuations in all the species of the Serengeti, brought about by such perturbations as disease (and its elimination) and climatic change. Thus, while each population goes through cycles of boom and bust, the system as a whole maintains its health and overall productivity. It is the diversity of species within the system that lends sustainability to the entire system.

The Masai are an ethnic group that have lived as pastoralists in and around the Serengeti, long before the beginning of colonialism in Africa. The earliest remains of domestic stock in East Africa date back to around 3500 years BP (Homewood and Rogers 1991). The grazing patterns of pastoralist herds were very similar to the grazing patterns of wildebeest. The close parallel between the ranging strategies of pastoralist herds and wildlife is dictated by their common dependence on critical grazing, mineral, and water resources.

The episodes for drought, fire, epidemic and herbivore population fluctuations of the region fostered the evolution of a risk-reducing nomadic system. Among the management techniques used by the traditional Masai were:

- assuring their herds had access to refuges from drought (usually highland or swamp pastures).

- maintaining a flexible mixture of stock species with different feeding, ranging, production, disease- and drought-resistance, and reproductive characteristics. Small stock were important in post-disaster herd reconstitution; large stock were the preferred investment, once a critical threshold of stock was reached.

- maintaining herds with a high proportion of females capable of rapid reproductive response in the aftermath of disaster.

- maximizing stock numbers in hope of retaining enough to ensure long-term survival of the herd, despite heavy periodic drought and disease losses.

- splitting the stock holdings into different herding units according to species, maturity and reproductive condition.

- maintaining social systems of stock loan and redistribution among friends and kinsmen that spread risk over a wider geographical area and range of habitat types — thus buffering disaster (Homewood and Rogers 1991, 143).

Between 1920 and 1960, the Masai were progressively evicted from their traditional grazing lands, with areas being set aside for expatriate settlers, cultivators from politically more dominant tribes, wildlife preserves, and major agribusiness and plantation schemes. To compensate the Masai for the prohibition against using the traditional commons, they were offered group ranches. The title to the demarcated areas was held by a group of pastoralist families who would jointly take out loans, plan and commission water developments and other improvements, control stocking rates, and cooperate on day-to-day grazing and marketing management. The Masai were transformed from a subsistence culture to one partially integrated with the national economy.

Despite the expenditure of tens of millions of dollars, the development projects were a disaster for the Masai. There was little or no consultation with Masai on geographical boundaries or specific aspects of development. Several ranching associations were set up, but only one achieved the promised title to the land. The project managed a fairly rapid installation of technical services like dips, dams, and other water development measures. This brought about massive uncontrolled immigration of both pastoralists and cultivators. Marketing arrangements originally set up under the control of ranching associations were brought back under state control. By 1978, the establishment of ranching associations was seen as conflicting with state goals, and no further ranching associations were granted occupancy rights.

By 1979, the dams, dips, and other technical inputs had deteriorated rapidly. Numerous postmortem reports evaluated the reasons for its failure. Lasting development had not occurred, and the Masailand infrastructure had dwindled since the 1950s. There was a decline in the availability of even the simplest introduced technologies, such as hand churns and hand grinders for maize (Homewood and Rogers 1991, 208).

Commenting on the efforts to transform grazing systems that have evolved to be sustainable in an unpredictable climate to those that conform to the demands of a modern economy, McNaughton (1993, 18) said, "Human management, even the most intelligent and enlightened, is not as effectual at facilitating species preservation at multiple trophic levels and maintaining sustained levels of productivity in grazing ecosystems as are the mechanisms produced by 50 million years of evolution, including coevolution, of grasses, the herbivores that feed on them, and other members of natural grassland trophic webs."

TRANSITION FROM UNSUSTAINABILITY TO SUSTAINABILITY

Agriculture

Tomé Açu, Brazil

In 1929, 43 families of Japanese immigrants arrived in Tomé-Açu, Brazil, about 100 miles south of Belém in the Eastern Amazonian rain forest (Sioli 1973). They were sponsored by a large Japanese textile company interested in the settlement of the Amazon region as a means of relieving population pressure in their home district in Japan. The colony initially tried to raise cacao, but this effort failed. Two of the reasons included the low soil fertility and fungal disease. The colony began to decline, and the process was accelerated by a severe malaria epidemic. The remaining colonists had to resort to shifting cultivation of rice, corn, beans, and manioc. However, they began to experiment with vegetable crops brought from Japan, including tomatoes, bell peppers, cucumbers, radishes, and turnips.

Efforts at developing a marketing cooperative were interrupted by World War II. Tomé-Açu was designated as an enemy alien relocation center, and transport of produce to Belém was halted. This provided them with time to restructure their cooperative. They discovered that black pepper, originally brought to the colony in 1930, could be profitably grown on the poor soils. However, as plantations of black pepper age, they become increasingly susceptible to the fungus *Fusarium solani*. It became necessary for the Japanese to integrate other crops into their production and marketing system (Jordan 1987).

Because of the low nutrient content of Amazonian soils, sustainability of production depends on maintaining soil fertility. This is achieved by a system of production in which the ground is laid bare for only one or two years during production of high value annuals. Following that, a mixture of perennial species recycles nutrients by means of leaf and litter fall. This litter forms a mulch on the soil surface that releases nutrients in synchrony with the demands of growing crops.

A cycle may start with the clearing of secondary forest growing on a site thought ready for cultivation (see photoessay "Agroforestry in the Tropics"). Some trees may be used for lumber, and others converted to charcoal. After the plot is cleared, remaining slash is burned, and seedlings of rubber trees tolerant of the low-fertility soils may be planted at intervals of several meters. For several years before the canopies of the rubber trees close, other crops are planted to take advantage of the freshly fallowed soil and to maintain soil cover.

First, corn is planted between the rubber seedlings. When the corn is half a meter tall, ginger is planted. When the corn is harvested, the stalks and leaves are collected, and this mulch is spread around the base of fruit tree seedlings in neighboring plantations to conserve moisture and improve soil quality. Cotton, winged beans, and peanuts are planted in the same plot following corn harvest. These crops do well because of the fertilizing effect of decaying corn roots. When the ginger is harvested, the husks — which the farmers claim repel nematodes — are spread on nearby beds of onions, which are susceptible to the nematodes.

After two years, the canopy of the rubber trees begins to close. A high-yield variety of rubber is grafted to the root stock, and later, a second, fungal-resistant graft may be added. Nonresistant leaves in

Amazon rubber plantations usually will be killed by the leaf fungus *Dothidella ulei*.

Pepper is still an important crop at Tomé Açu. It can be cultivated for almost ten years before problems with rot force abandonment. Pepper is a vine, and living trees of nitrogen-fixing species are sometimes used to support the vines. Palm species valuable for heart of palm, fruit used for juice, and for use as pig food in neighboring farms are often seeded in between the pepper plants by parrots and macaws, and these palms may be left to grow. Several years before a pepper plantation is abandoned, tree species with high commercial value for wood may be planted among the pepper vines. When the pepper plantation is abandoned, the plot is already supporting a vigorously growing forest of valuable species.

Cacao is also an important crop. Disease in cacao plantations is controlled by pruning branches every year, which increases wind flow and decreases humidity. In some plantations, vanilla (an extremely high-value vine) is planted at the base of cacao trees, which provide the support.

Depending upon market conditions, the exact sequence of species planted may vary. However, the temporal sequence of structure and function varies little. It begins with annuals and short-lived perennials and progresses to long-lived perennials, with an accompanying change in structure from open, bushy or vine-like vegetation to a closed, multilayered forest. The sequence is similar to the natural succession that occurs spontaneously in abandoned clearings.

The Japanese cropping systems generally require large inputs of fertilizers. Although some fertilizers are purchased, there are many examples of the use of organic materials derived from within the farms. Cacao hulls are often composted or burned and returned to cacao plantations. Weeds are normally cut and left in place or piled as mulches. Some farmers make use of green manures such as nitrogen-fixing leguminous herbs, shrubs and trees. Rice farmers trade rice husks to chicken farmers who spread the hulls on the floors of chicken coops. After the hulls absorb the nutrient-rich poultry manure, they are bagged and traded to vegetable or fruit producers. As crop systems mature toward the tree-based stages, there is a trend toward the reduction in quantities of fertilizers applied. The farmers believe that the trees improve soil fertility.

Capital to finance farm activities comes from the community cooperative. This cooperative has played a central role in the development and success of the agricultural community. The cooperative provides easy access to materials and equipment, credit, technical assistance, and local processing of farm products. It plays an indispensable role in marketing of crops, as well as the development of markets for new crops. It also serves as an important center for information exchange between local farmers and farmers from other regions. Through the cooperative, the farmers are able to assert leverage in the marketplace (Subler and Uhl 1990).

This type of sustained management is unusual in the Amazon. The complexity of management and the level of technical expertise that the Japanese have obtained through years of experience and research represent an obstacle to untrained small producers. Another factor that may be important in the economic success of Tomé-Açu is cultural discipline. The work is often tedious and requires a type of discipline that occurs in traditional Japanese society, but is not common in other cultures.

Tropical Forestry

Reduced-Impact Logging

The tropical shelterwood forestry system was a sustainable management system in that it ensured that there was adequate reproduction of desirable species before logging took place, and care was exercised not to damage the seedlings and saplings. The system was abandoned, however, because of economic pressure. Greater economic gain could be obtained by clear-cutting the forest and planting agricultural crops.

Recently, however, a change in economic incentives has spurred interest in reviving sustainable tropical forestry, under the name of "Reduced-Impact Logging." A reduced impact logging trial began in 1992, when New England Electric System of Massachusetts provided funds to Innoprise Corporation Sendirian Berhad of Sabah, Malaysia, to implement a set of reduced-impact logging guidelines in 1400 hectares of tropical forest (Pinard *et al.* 1995). An important motivation has been the possibility of a carbon tax or mandatory reductions in greenhouse emissions in the United States being imposed. The United Nations Conference on the Environment and Development (1992) and the U.S. Energy Policy Act suggest that such programs may soon be developed. If the U.S. government mandates emissions reductions, and if policies develop that accept reductions in net emissions (whether domestic or international), the carbon retained in the forest due to the company's investment may be credited toward taxes on its greenhouse gas emissions.

Reduced Harvesting Guidelines include specifications for pre-harvest planning, vine cutting, felling, skidding, and post-harvest site closure (Uhl *et al.* 1997).

- **Planning**. Preharvest planning requires a stock map of harvestable trees. The map also shows stream and road buffer zones, and sensitive areas to be excluded from logging. Roads and skid trails generally are located on ridges to avoid steep grades, facilitate uphill skidding, minimize skidding distances and stream crossings, and reduce the amount of soil eroded into streams. Main extraction routes and landing areas are located on the map and marked in the field with paint. Rangers mark trees to be felled with a number and a vertical paint blaze to indicate the intended direction of fall.

- **Vine cutting**. Canopies of tropical forest trees are linked together with vines. As a result, when one tree is cut, many are damaged or pulled down. To eliminate this domino effect, vines are cut about one year prior to logging. Another advantage of vine cutting is the increased light that reaches the understory, possibly initiating physiological changes in tree seedlings that will be of benefit following logging.

- **Tree felling**. Direction of tree fall is guided by considerations about safety, ease of skidding, avoidance of damage to remaining trees, and effects on buffer zones. Because of crown irregularities, it is sometimes difficult to judge precisely the direction in which a tree will fall. When there is question about potential damage in the reduced impact logging system, a tree may not be marked.

- **Winching and Skidding**. Main skid trails are shown on the stock map (bulldozers are restricted to these main trails). The scraping of surface soil is restricted as much as possible. Winches are used to move logs from stumps to the main skid trails. Because soils are more susceptible to erosion during wet weather, harvest is restricted to dry weather.

- **Logging area closure.** After logging is completed in a harvest unit, the skid trails are closed, and cross-drains are installed at intervals on slopes, to minimize erosion.

Range Management

The key to sustainable range management is to rotate the herd continually between different areas of the pasture or range. The length of time depends upon many factors, including the species of grazer, and the rate at which other areas recover. A naturally sustainable grazing system existed in the Serengeti. There, herds of wild grazers migrated throughout a rangeland during the course of a year following the new growth of grass, and leaving the old to recover. In the United States, the bison herds probably followed circular migratory routes, perhaps in response to available water and feed, and might have grazed in about the same place at the same time every year (Heady and Child 1994).

Sustainable Range Grazing

There are infinite combinations of grazing patterns, each suited to a particular species of grazer, to the local environment, and to the economy. The "Hema System" in the Middle East is an example. It predates the Islamic era in Saudi Arabia, Syria, Lebanon, Tunisia, and other countries in the Middle East (Heady and Child 1994). In this system, animal grazing is rotated throughout a range. The rotation depends not only upon weather, but also upon the use of land for other purposes, such as bee keeping and growing firewood.

Fire is often an important factor in rotational grazing schemes, as they burn back woody invaders and stimulate the growth of grass. Timing, however, is important; in order to gain maximum impact from the burn, it should occur when undesirable species are particularly sensitive. For example, in regions where summer grasses are important, spring burns are more effective than winter burns, because the fire will damage woody invaders more than it will the native grasses.

An important factor in range and pasture sustainability is the carrying capacity — that is, the number of animals that can be continually supported. Prolonged trampling and overgrazing of pastures or rangeland has adverse effects, such as pulverization of the soil surface, excessive soil compaction, and injury to plants.

Overgrazing is the most common cause of range and pasture deterioration. However, the number of animals on a pasture may not be as important as the length of time they are allowed to remain in an area (Savory 1988). For example, a herd of 100 cows on a 10 acre pasture for a whole year could easily destroy the productive capacity of that pasture. However, the same traffic produced by 100 cattle on a single day, followed by 364 days of recovery time, would produce a far different result. The plant and soil communities could recover from the damaging component of the trampling and benefit from the intense deposition of dung and urine.

Maximum impact over minimum time, followed by a sufficient recovery period, makes trampling an extremely effective tool for maintaining brittle rangeland and watersheds. To maintain healthy, productive, diverse perennial grassland in brittle environments, grazers should periodically remove old plant parts and scatter them on the ground, break the soil surface crust and algal communities between plants, and provide some compaction to the soil to allow grass seedlings to establish (Savory 1988).

Feedlot Production

The most important difference between range grazing and modern feedlot production is that in range grazing, the animals are allowed to perform many of the services of nature that lend sustainability to rangeland. When animals are confined to pens or small areas, their "services" are so concentrated that they become damaging. For example, in feedlot production, manure and urine contaminate ground water and streams. Under rotational management, manure and urine contribute to soil fertility and the sustainability of forage production.

Feedlot production replaces grazing on the open range because it can be more economically profitable. The cost of keeping livestock in pens and transporting food to them is more than compensated for by the increased rate at which livestock gain weight while in pens. The cost of petroleum expended in cultivating grain and hauling it to feed lots is less than the increased profits generated by not having the livestock expend energy on the open range.

FACTORS CAUSING TRANSITIONS

It is sometimes assumed that resources are managed unsustainably because knowledge is lacking on techniques for sustainable management. The lesson of this chapter is that this is rarely the case. In many instances, factors other than lack of technical knowledge are responsible for unsustainable management.

In rice cultivation in the Sahel by the Marka and in grazing systems of the Masai, sustainable resource management evolved over centuries. Indigenous peoples were keenly aware of the delicate balance between sustainability and productivity, and they fine-tuned their management systems to ensure an optimal balance. The key element was an awareness of the very site-specific relationship between the environment and the variety of rice (in the case of the Marka), or between the season and the migration of herds (in the case of the Masai). In the Marka situation, sustainability was curtailed by an attempt to homogenize the resource and the production system, so as to exploit the product for greater short-term gain and to integrate the producers into the national and international economy. The Marka system ended when the goal of "regional development" provided perverse incentives that encouraged the local people to abandon their traditional knowledge.

The Masai system ended when traditional forces of colonialism expropriated the range, divided it up, and fenced it off. This action severed the interaction between the grazers and the range, an interaction critical to the survival of both.

In the case of Tome Açu, the reverse happened. There was a gradual incorporation of the lessons of nature into the cultural system. The early Japanese colonists began with an agricultural system that resembled a degrading type of slash-and-burn cultivation. However, because of their restriction to a limited area of land, they had to develop a system of agriculture that would not deplete that land. As a result of several generations of empirical experimentation, they evolved a system that closely resembled the natural succession of plant communities in the region.

In the case of tropical forestry, economic factors have caused the failure of sustainable management systems. Harvest to ensure reproduction of the forest requires time and effort that add to the cost of the final product. Because there is always someone willing to log unscrupulously in virgin stands of timber when access is subsidized by government sponsored roads or concessions, sustainable forestry will never be able to compete until opportunities for reckless logging are eliminated, or sustainable practices are rewarded.

The case study of Jari illustrates the strength of cultural and economic dominance in resource management decisions. In this particular case, industrial forestry was implemented in the Amazon, because it was assumed that such an approach was superior. Despite overwhelming evidence that the effort was a disaster, proponents of the scheme still could not admit that the technological and capital intensive approach was inferior. Even in the face of ecological degradation, social upheaval, and tremendous economic loss, it was still heralded as a success (Hornick *et al.* 1984), and has served as a classic example of the arrogance of power.

CHAPTER 11
CONCLUSION

Throughout most of history, humans have had to struggle against nature to survive. Nature was a formidable opponent, and unless humans adapted to nature, nature would exterminate them. Only very recently has the belief arisen that humans have mastered the power to conquer and regulate nature, and become freed from the threat of scarcities or disease by ominous forces outside their control.

Is this belief justified? Possibly. But up to the present time, there is enough evidence to suggest that increasing dependence upon conquering nature may not be the wisest long-term strategy. The more that technology is applied to control nature, the greater the negative feedback from nature to humans, and the more technology that must be applied to keep the planet livable.

It may be worthwhile to consider a different approach. Rather than increasing attempts to control nature, it may be beneficial to try to understand and cooperate with her, instead of continually fighting her.

Much is already known about how to manage plants, animals, and soil in a way that takes advantage of the many "free" services of nature, such as nutrient recycling and disease suppression. However, there is still much to be learned about managing for sustainability, to implement what we already know, and to integrate sustainability into the market system.

Why is not more effort being dedicated to cooperating with nature? Why are not the big land grant universities funding more studies on how to manage sustainably? Why are the big agribusinesses not dedicating more effort to understanding nature instead of continuing to mold nature into a corporate image?

The answer is short-term economics. It is more economically profitable in the short term to conquer than to cooperate. The tool of conquest — petroleum — yields profits more quickly than the tool of cooperation — site-specific human knowledge. Until it becomes more profitable to cooperate, there will be little significant change. Sustainable management cannot compete in the short-term free market with management based on exploitation, or management subsidized to maximize short-term productivity.

The first step in the transition to sustainability is to give farmers an incentive to change the way that they manage farms, the foresters the way they manage woodlands, and the ranchers the way they manage grasslands. Just as resource management had to be subsidized as the United States was settled in order to produce enough food and fiber to support a growing society, sustainable resource management must now be subsidized, at least initially, so that such approaches can become economically competitive.

A related economic problem is the resistance to change by agribusiness and big timber companies that have invested billions of dollars in traditional methods of resource exploitation. Changing methods of resource management threatens these companies and their shareholders. These companies and their lobbyists in the nation's capital will fight to convince lawmakers to leave things as they are. Like any well-organized minority, they are able to impose upon an unorganized majority. However, there is hope. Many of the nongovernmental organizations in the United States that once were focused upon nature preservation are now assuming a more active role in promoting research and development of sustainable management, as well as lobbying for its adoption.

What can be said to those who argue that we cannot afford to cut back on the rate of agricultural expansion? In order to feed the swelling world population and to satisfy the increasing material standards of living of those in developed countries, they say it will be necessary to focus all efforts on maximizing productivity, and increasing the amount of land dedicated to agriculture.

The answer is that the problem is not our ability to produce food and fiber — the problem is our desire to produce it cheaply. A recent story in the *Wall Street Journal* (Zachary 1996) said that the prices of food commodities are so low that they comprise only two percent of all global costs. "The grain that goes into a box of breakfast cereal," the article stated, "costs less than the paper for the box." Now, if producing grain sustainably adds ten percent or fifty percent, or even one hundred percent to the cost of producing it unsustainably, would it not be worth it? Would not it seem reasonable to value the cereal at least as much as the box?

Further, many of the techniques suggested to increase sustainability could result in production that is as high as petroleum-subsidized agriculture. (At present they could not, but that is because virtually no effort is being made in that direction.) Land grant colleges of agriculture and forestry dedicate little effort to research in agroforestry, agroecology, and other approaches toward working with nature. Their focus remains primarily on chemical and genetic technology, and the philosophy of adapting production systems to match the demands of the economic market system. Only when the public begins to understand that sustainability depends upon lessening our dependence upon artificial subsidies and making our economic system adjust to nature will the resource managers begin to respond.

We cannot, of course, abandon all technology and revert to rice cultivation along the lines of the traditional Marka in Africa. But we can begin to reverse the trend of the tightening ratchet.

Literature Cited

Ae, N. J., J. Arihara, K. Okada, T. Yoshihara, and C. Johansen. 1990. Phosphorus uptake by pigeon pea and its role in cropping systems of the Indian subcontinent. *Science* 248:477–480.

Alegre, J. C., and M. R. Rao. 1996. Soil and water conservation by contour hedging in the humid tropics of Peru. *Agriculture, Ecosystems and Environment* 57:17–25.

Alstad, D. N., and D. A. Andow. 1995. Managing the evolution of insect resistance to transgenic plants. *Science* 268:1894–1896.

Altieri, M. A. 1987. Pest management. In *Agroecology: The Scientific Basis of Alternative Agriculture*, ed. M. A. Altieri, 159–172. Boulder, CO: Westview Press.

Altieri, M. A., ed. 1995. *Agroecology: The Science of Sustainable Agriculture*. Boulder, CO: Westview Press.

Altieri, M. A., and M. Liebman. 1986. Insect, weed, and plant disease management in multiple cropping systems. In *Multiple Cropping Systems*, ed. C. A. Francis. 183–218. New York: Macmillan.

Altieri, M. A., and L. L. Schmidt. 1985. Cover crop manipulation in northern California orchards and vineyards: effects on arthropod communities. *Biological Agriculture and Horticulture* 3:1–24.

Altieri, M. A., D. L. Glaser, and L. L. Schmidt. 1990. Diversification of agroecosystems for insect pest regulation: experiments with collards. In *Agroecology: Researching the Ecological Basis for Sustainable Agriculture*, ed. S. R. Gliessman, 70–82. New York: Springer-Verlag.

Andow, D. 1983. The extent of monoculture and its effects on insect pest populations with particular reference to wheat and cotton. *Agriculture, Ecosystems and Environment* 9:25–36.

Annis, P. C., and C. J. Waterford. 1996. Alternatives — Chemicals. In *The Methyl Bromide Issue*, eds. C. H. Bell, N. Price, and B. Chakrabarti, 275–322. New York: Wiley.

Arnold, S. F., D. M. Klotz, B. M. Collins, P. M. Vonier, L. J. Guillette, and J. A. McLachlan. 1996. Synergistic activation of estrogen receptor with combinations of environmental chemicals. *Science* 272:1489–1492.

Athari, S., and H. Kramer. 1983. The problem of determining growth losses in Norway spruce stands caused by environmental factors. In *Effects of Accumulation of Air Pollutants in Forest Ecosystems*, eds. B. Ulrich and J. Pankrath, 319–325. Holland: Dordrecht.

Avery, D. T. 1995. *Saving the Planet with Pesticides and Plastic*. Indianapolis, Indiana: Hudson Institute.

Ba, T. A., and B. Crousse. 1985. Food-production systems in the Middle Valley of the Senegal River. *International Social Science Journal* 37(3):389–400.

Barrett, G. W., N. Rodenhouse, and P. J. Bohlen. 1990. Role of Sustainable Agriculture in Rural Landscapes. In *Sustainable Agricultural Systems*, eds. C. A. Edwards, L. Rattan, P. Madden, R. H. Miller, and G. House, 624–636. Delray Beach, FL: St. Lucie Press.

Barrett, S. A. 1925. *Cayapa Indians of Ecuador*. Indian Notes and Monographs No. 10, New York Museum of the American Indian. New York: Heye Foundation.

Batmanian, G. 1990. Reforestation of degraded pastures in the Brazilian Amazon: effect of site preparation on phosphorus availability in the soil. Ph.D. dissertation, University of Georgia, Athens.

Baumgartner, A., and M. Kirchner. 1980. Impacts Due to Deforestation. In *Interactions of Energy and Climate*, eds. W. Bach, J. Pankrath, and J. Williams, 305–316. Dordrecht: Reidel.

Beissinger, S. R., and E. H. Bucher. 1992. Can parrots be conserved through sustainable harvesting? *BioScience* 42:164–173.

Benenson, A. S., ed. 1995. *Control of Communicable Diseases Manual*. Washington, DC: American Public Health Association.

Bhat, B. K. 1995. Breeding methodologies applicable to pyrethrum. In *Pyrethrum Flowers*, eds. J. E. Casida and G. B. Quistad, 67–94. New York: Oxford Univ. Press.

Boucher, D. H. 1985. Mutualism in Agriculture. In *The Biology of Mutualism*, ed. D. H. Boucher, 375–386. New York: Oxford Univ. Press.

Brady, N. C. 1974. *The Nature and Properties of Soils*. New York: Macmillan.

Browder, J. O. 1992. The limits of extractivism. *BioScience* 42:174–182.

Buckley, J. 1986. Environmental Effects of DDT. In *Ecological Knowledge and Environmental Problem Solving*, 358–371. Washington, DC: National Academy Press.

Burkart, M. R., and D. W. Kolpin. 1993. Hydrologic and land-use factors associated with herbicides and nitrate in near-surface aquifers. *J. Environmental Quality* 22:646.

Carroll, C. R., and S. J. Risch. 1989. An evaluation of ants as possible candidates for biological control in tropical annual systems. In *Agroecology*, ed. S. R. Gleisman, 30–46. New York: Springer-Verlag.

Chapin, F. S. 1983. Pattern of Nutrient Absorption and Use by Plants from Natural and Man-Modified Environments. In *Disturbance and Ecosystems*, eds. H. A. Mooney and M. Godron, 175–187. Berlin: Springer-Verlag.

Chapin, F. S., B. H. Walker, R. J. Hobbs, D. U. Hooper, J. H. Lawton, O. E. Sala, and D. Tilman. 1997. Biotic control over the functioning of ecosystems. *Science* 277:500–504.

Chatterton, L., and B. Chatterton. 1996. *Sustainable Dryland Farming*. Cambridge: Cambridge Univ. Press.

Cheshire, M. V. 1985. Carbohydrates in relation to soil fertility. In *Soil Organic Matter and Biological Activity*, eds. D. Vaughan and R. E. Malcolm, 263–288. Dordrecht: Junk.

Chou, C. H. 1990. The role of allelopathy in agroecosystems: studies from tropical Taiwan. In *Agroecology*, ed. S. R. Gliessman, 104–121. New York: Springer-Verlag.

Clark, J. B. 1994. Genetic engineering ("Biotech"): use of science gone wrong. In *Agricultural Biotechnology and the Public Good*, ed. J. F. MacDonald, 191–193. Ithaca, NY: National Agricultural Biotechnology Council.

Clusener-Godt, M., and I. Sachs. 1994. Extractivism in the Brazilian Amazon: Perspectives on Regional Development. *MAB Digest* 18. Paris: UNESCO.

Cohen, J. E., and D. Tilman. 1996. Biosphere 2 and biodiversity: the lessons so far. *Science* 274:1150–1151.

Colborn, T. D. Dumanoski, and J. P. Myers. 1996. *Our Stolen Future*. New York: Dutton.

Coleman, D. C., and D. A. Crossley Jr. 1996. *Fundamentals of Soil Ecology*. San Diego: Academic Press.

Coleman, D. C., P. F. Hendrix, M. H. Beare, D. A. Crossley, S. Hu, and P. C. J. van Vliet. 1994. The impacts of management and biota on nutrient dynamics and soil structure in sub-tropical agroecosystems: impacts on detritus food webs. In *Soil Biota: Management in Sustainable Farming Systems*, eds. C. F. Pankhurst, B. M. Doube, V. V. S. R. Gupta, and P. R. Grace, 133–143. East Melbourne: CSIRO.

Cook, R.J. 1994. Introduction of soil organisms to control root diseases. In *Soil Biota: Management in Sustainable Farming Systems*, eds. C. F. Pankhurst, B. M. Doube, V. V. S. R. Gupta, and P. R. Grace, 13–22. East Melbourne: CSIRO.

Culotta, E. 1996. Exploring biodiversity's benefits. *Science* 273:1045–1046.

Cussans, G. W. 1995. Integrated weed management. In *Ecology and Integrated Farming Systems*, eds. D. G. Glen, M. P. Graves, and H. M. Anderson, 17–29. Chichester: Wiley.

DeBlieu, J. 1992. Could the red wolf be a mutt? *New York Times Magazine*, 14 June, 30–46.

del Moral, R., and C. H. Muller. 1970. The allelopathic effects of *Eucalyptus camaldulensis*. *American Midland Naturalist* 83:254–282.

Desowitz, R. S. 1987. *New Guinea Tapeworms and Jewish Grandmothers: Tales of Parasites and People*. New York: Norton.

Dubois, J. C. L. 1996. Uses of wood and non-wood forest products by Amazon forest dwellers. *Unasylva* 186:8–15.

Erdle, T. A., and G. L. Baskerville. 1986. Optimizing timber yields in New Brunswick forests. In *Ecological Knowledge and Environmental Problem–Solving*, 275–300. Washington, DC: National Academy Press.

Evans, J. 1992. *Plantation Forestry in the Tropics*. Oxford: Clarendon Press.

Ewel, J., F. Benedict, C. Berish, and B. Brown. 1982. Leaf area, light transmission, roots, and leaf damage in nine tropical plant communities. *Agro-ecosystems* 7:305–326.

Federici, B. A., and J. V. Maddox. 1996. Host specificity in microbe–insect interactions. *BioScience* 46:410–421.

Ferguson, B. K. 1983. Whither water? The fragile future of the world's most important resource. *The Futurist*, April, 29–47. Cited in Soule, J., D. Carre, and W. Jackson. 1990. Ecological impact of modern agriculture. In *Agroecology*, eds. C. R. Carroll, J. H. Vandermeer, and P. Rosset, 165–188. New York: McGraw-Hill.

Fineblum, W. L., and M. D. Rausher. 1995. Tradeoff between resistance and tolerance to herbivore damage in a morning glory. *Nature* 377:517–520.

Folke, C., M. Hammer, R. Costanza, and A. Jannson. 1994. Investing in natural capital — why, what, and how? In *Investing in Natural Capital: The Ecological Economics Approach to Sustainability*, eds. A. Jannson, M. Hammer, C. Folke, and R. Constanza, 1–20. Covelo, CA: Island Press.

Fox, J. E. D. 1976. Constraints on the natural regeneration of tropical moist forest. *Forest Ecology and Management* 1:37–65.

Francis, C. A. 1986. *Multiple Cropping Systems*. New York: Macmillan.

Francis, C. A. 1989. Biological efficiencies in multiple-cropping systems. *Advances in Agronomy* 42:1–42.

Fry, W. E., and S. B. Goodwin. 1997. Resurgence of the Irish potato famine fungus. *BioScience* 47:363–371.

Fulhage, C. 1994. Livestock manure systems trends in the Midwest. In *Liquid Manure Application Systems: Design, Management and Environmental Assessment*, 21–24. Proceedings from a Conference, Rochester, New York. Ithaca, NY: N.E. Agricultural Engineering Service.

Gajaseni, J. 1992. Overview of taungya. In *Taungya: Forest Plantations with Agriculture in Southeast Asia*, eds. C. F. Jordan, J. Gajaseni, and H. Watanabe, 3–8. Wallingford, UK: CAB International.

Gajaseni, J., and C. F. Jordan. 1990. Decline of teak yield in Northern Thailand: effects of selective logging on forest structure. *Biotropica* 22:114–118.

Gajaseni, J., and C. F. Jordan. 1992. Theoretical basis for taungya and its improvement. In *Taungya: Forest Plantations with Agriculture in Southeast Asia*, eds. C. F. Jordan, J. Gajaseni, and H. Watanabe, 68–81. Wallingford, UK: CAB International.

Gentry, A., and C. H. Dodson. 1991. Biological extinction in Western Ecuador. *Annals of the Missouri Botanical Garden* 78:273–295.

Gerwing, J. J., J. S. Johns, and E. Vidal. 1996. Reducing waste during logging and log processing: forest conservation in eastern Amazonia. *Unasylva 187*, 47:17–25.

Gillette, D. 1994. Common environmental problems arising from liquid manure systems. In *Liquid Manure Application Systems: Design, Management and Environmental Assessment*, 126–138. Proceedings from a Conference, Rochester, New York. Ithaca, NY: N.E. Agricultural Engineering Service.

Gliessman, S. R. 1983. Allelopathic interactions in crop–weed mixtures. *Journal of Chemical Ecology* 9:991–999.

Gordon, J. C. 1996. Trends and developments in modern forestry. In *The Literature of Forestry and Agroforestry*, eds. P. McDonald and J. Lassoie, 1–14. Ithaca, NY: Cornell Univ. Press.

Grace, P. R., J. N. Ladd, and J. O. Skjemstad. 1994. The effect of management practices on soil organic matter dynamics. In *Soil Biota: Management in Sustainable Farming Systems*, eds. C. F. Pankhurst, B. M. Doube, V. V. S. R. Gupta, and P. R. Grace, 162–171. East Melbourne: CSIRO.

Guedes, R. 1993. Phosphorus mobilization by root exudates of pigeon pea (*Cajanus cajan*). Ph.D. dissertation, University of Georgia, Athens.

Gupta, V. V. S. R. 1994. The impact of soil and crop management practices on the dynamics of soil microfauna and mesofauna. In *Soil Biota: Management in Sustainable Farming Systems*, eds. C. F. Pankhurst, B. M. Doube, V. V. S. R. Gupta, and P. R. Grace, 107–132. East Melbourne: CSIRO.

Halstead, P., and O'Shea, J. 1989. Introduction: cultural responses to risk and uncertainty. In *Bad Year Economics: Cultural Responses to Risk and Uncertainty*, eds. P. Halstead and J. O'Shea, 1–7. New York: Cambridge Univ. Press.

Halvorson, W. L., and G. E. Davis, eds. 1996. *Science and Ecosystem Management in the National Parks*. Tucson: Univ. of Arizona Press.

Harmon, M. E., W. K. Ferrell, and J. F. Franklin. 1990. Effects on carbon storage of conversion of old-growth forests to young forests. *Science* 247:699–702.

Harrison, H. L., O. L. Loucks, J. W. Mitchell, D. F. Parkhurst, C. R. Tracy, D. G. Watts, and V. J. Yannacone. 1970. Systems studies of DDT transport. *Science* 170:503–512.

Hart, R. D. 1980. A natural ecosystem analog approach to the design of a successional crop system for tropical forest environments. *Biotropica*, 12(suppl.):73–82.

Hartshorn, G. S. 1990. Natural forest management by the Yanesha Forestry Cooperative in Peruvian Amazonia. In *Alternatives to Deforestation: Steps Toward Sustainable Use of the Amazon Rain Forest*, ed. A. B. Anderson, 128–151. New York: Columbia Univ. Press.

Hasan, S., E. S. DelFosse, E. Aracil, and R. C. Lewis. 1992. Host-specificity of *Uromyces heliotropii*, a fungal agent for the biological control of common heliotrope (*Heliotropium europaeum*) in Australia. *Annals of Applied Biology* 121:697–705.

Hasse, V., and J. A. Litsinger. 1981. The influence of vegetational diversity on host finding and larval survivorship of the Asian corn borer, *Ostrinia furnacalis*. IRRI Saturday Seminar, Entomology Dept., International Rice Research Institute, The Philippines. Cited in Francis, C. A. 1986. *Multiple Cropping Systems*, 186. New York: Macmillan.

Heady, H. F., and R. D. Child. 1994. *Rangeland Ecology and Management*. Boulder, CO: Westview Press.

Heitz, J. R. 1995. Pesticidal applications of photoactivated molecules. In *Light–Activated Pest Control*, eds. J. R. Heitz and K. R. Downum, 1–16. Washington, DC: American Chemical Society.

Henshel, D. S., J. W. Martin, and J. C. Dewitt. 1997. Brain asymmetry as a potential biomarker for developmental TCDD intoxication: a dose–response study. *Environmental Health Perspectives* 105(7): 718–725.

Hoffman, M. P., C. H. Petzoldt, C. R. MacNeil, J. J. Mishanec, M. S. Orfanedes, and D. H. Young. 1995. Evaluation of an onion thrips pest management program for onions in New York. *Agriculture, Ecosystems and Environment* 55:51–60.

Homewood, K. M., and W. A. Rogers. 1991. *Maasailand Ecology: Pastoralist development and wildlife conservation in Ngorongoro, Tanzania.* Cambridge: Cambridge Univ. Press.

Hornick, J. R., J. I. Zerbe, and J. L. Whitmore. 1984. Jari's successes. *Journal of Forestry* 82(11):663–670.

Hughes, G., and J. L. Gonzalez Andujar. 1997. Simple rules with complex outcomes. *Nature* 387:241–242.

Huo, Y. 1992. Taungya in China. In *Taungya: Forest Plantations with Agriculture in Southeast Asia*, eds. C. F. Jordan, J. Gajaseni, and H. Watanabe, 133–146. Wallingford, UK: CAB International.

Iltis, H. H., J. F. Doebley, R. M. Guzmán, and B. Pazy. 1979. Zea diploperennis (Gramineae), a new teosinite from Mexico. *Science* 203:186–188.

International Society of Tropical Forestry. 1985. Jari among top 100 agribusiness firms. *Inter. Soc. Trop. Forestry News* 6:7.

Jackson, W. 1985. *New Roots for Agriculture.* Lincoln: Univ. of Nebraska Press.

Janzen, D. H. 1985. The natural history of mutualisms. In *The Biology of Mutualisms*, ed. D. H. Boucher, 40–99. New York: Oxford Univ. Press.

Jordan, C. F. 1982. The nutrient balance of an Amazonian rain forest. *Ecology* 63:647–654.

Jordan, C. F. 1985. Jari: a development project for pulp in the Brazilian Amazon. *The Environmental Professional* 7:135–142.

Jordan, C. F. 1985. *Nutrient Cycling in Tropical Forest Ecosystems.* Chichester: Wiley.

Jordan, C. F. 1987. Agroecology at Tomé-Açu, Brazil. In *Amazon Rain Forests: Ecosystem Disturbance and Recovery*, ed. C. F. Jordan, 70–73. New York: Springer-Verlag.

Jordan, C. F. 1989. *An Amazon Rain Forest: The Structure and Function of a Nutrient Stressed Ecosystem and the Impact of Slash and Burn Agriculture.* Paris: UNESCO, and Carnforth, UK: Parthenon Publishing.

Jordan, C. F. 1995. *Conservation: Replacing Quantity with Quality as a Goal for Global Management.* New York: Wiley.

Jordan, C. F., and E. G. Farnworth. 1982. Natural vs. plantation forests: a case study of land reclamation strategies for the humid tropics. *Environmental Management* 6:485–492.

Jordan, C. F., and C. E. Russell. 1989. Jari: A pulp plantation in the Brazilian Amazon. *Geo-Journal* 19.4:429–435.

Kadir, A. A. S. A., and H. S. Barlow. 1992. *Pest Management and the Environment in 2000*. Wallingford, UK: C.A.B. International.

Kaiser, J. 1996. New yeast study finds strength in numbers. *Science* 272:1418.

Kang, B. T., L. Reynolds, and A. N. Atta-Krah. 1990. Alley farming. *Advances in Agronomy* 43:315–359.

Kang, B. T., S. Hauser, B. Vanlauwe, N. Sanginga, and A. N. Atta-Krah. 1995. Alley farming research on high base status soils. In *Alley Framing Research and Development. International Institute of Tropical Agriculture*, eds. B. T. Kang, A. O. Asiname, and A. Larbi, 25–39. Nigeria: Ibadan.

Kass, D. L. 1997. Review of *Agroecology: The Science of Sustainable Agriculture*. *Agroforestry Systems* 35:111–115.

Keeler, K. H., C. E. Turner, and M. R. Bolick. 1996. Movement of crop transgenes into wild plants. In *Herbicide-Resistant Crops*, ed. S. O. Duke, 303–329. Boca Raton, FL: CRC Press.

Keever, C. 1950. Causes of succession on old fields of the Piedmont, North Carolina. *Ecological Monographs* 20:231–250.

Kjaergaard, T. 1995. Agricultural development and nitrogen supply from an historical point of view. In *Nitrogen Leaching in Ecological Agriculture*, ed. L. Kristensen, 3–14. Oxon, UK: Academic Publishers.

Kloepper, J. W. 1996. Host specificity in microbe–microbe interactions. *BioScience* 46:406–408.

Krimsky, S., and R. P. Wrubel. 1996. *Agricultural Biotechnology and the Environment*. Urbana: Univ. of Illinois Press.

Lavelle, P. 1996. Mutualism and soil processes: A Gaian outlook. In *Gaia in Action*, ed. P. Bunyard, 204–219. Edinburgh: Floris Books.

Leaky, R. 1996. Definition of agroforestry revisited. *Agroforestry Today* 8(1):5–7.

Lewis, D. H. 1985. Symbiosis and mutualism: crisp concepts and soggy semantics. In *The Biology of Mutualism*, ed. D. H. Boucher, 29–39. New York: Oxford Univ. Press.

Liebman, M. 1995. Polyculture cropping systems. In *Agroecology: The Science of Sustainable Agriculture*, ed. M. A. Altieri, 205–218. Boulder, CO: Westview Press.

Line, L. 1996. Lethal migration. *Audobon* 98(5):50–95.

Longley, M., and N. W. Sotherton. 1997. Factors determining the effects of pesticides upon butterflies inhabiting arable farmland. *Agriculture, Ecosystems and Environment* 61:1–12.

Lotz, L. A. P., J. Wallinga, and M. J. Kropff. 1995. Crop–weed interactions: quantification and prediction. In *Ecology and Integrated Farming Systems*, eds. D. G. Glen, M. P. Graves, and H. M. Anderson, 31–47. Chichester: Wiley.

Lovelock, J. 1996. The Gaia hypothesis. In *Gaia in Action*, ed. P. Bunyard, 15–33. Edinburgh: Floris Books.

Lowe, R. G. 1977. Experience with the tropical shelterwood system of regeneration in natural forest in Nigeria. *Forest Ecology and Management* 1:193–212.

Luck, R. F. 1986. Biological control of California red-scale. In *Ecological Knowledge and Environmental Problem Solving*, 165–189. Washington, DC: National Academy Press.

Macilwain, C. 1996. Bollworms chew hole in gene-engineered cotton. *Nature* 382:289.

MacKenzie, W. R., N. J. Hoxie, M. E. Proctor, M. S. Gradus, K. A. Blair, D. E. Peterson, J. J. Kazmierczak, D. G. Addiss, K. R. Fox, J. B. Rose, and J. P. Davis. 1994. A massive outbreak in Milwaukee of Cryptosporidium infection transmitted through the public water supply. *New England Jour. Medicine* 33:161–167.

Margulis, L. 1996. Jim Lovelock's Gaia. In *Gaia in Action*, ed. P. Bunyard, 54–64. Edinburgh: Floris Books.

Martin, D. L., and G. Gershuny. 1992. *The Rodale Book of Composting*. Emmaus, PA: Rodale Press.

Matta Machado, R. P., and C. F. Jordan 1995. Nutrient dynamics during the first three years of an alley cropping agroecosystem in southern USA. *Agroforestry Systems* 30:351–362.

McEvoy, P. B. 1996. Host specificity and biological pest control. *BioScience* 46:401–405.

McGaughey, W. H., and M. E. Whalon. 1992. Managing insect resistance to *Bacillus thuringiensis* toxins. *Science* 258:1451–1455.

McGuire, R. T. 1994. Liquid manure application systems: a public policy perspective. In *Liquid Manure Application Systems: Design, Management and Environmental Assessment*, 1–5. Proceedings from a Conference, Rochester, New York. Ithaca, NY: N.E. Agricultural Engineering Service.

McIntosh, R. J. 1993. The pulse model: Genesis and accommodation of specialization in the Middle Niger. *Journal of African History* 34:181–220.

McLachlan, J. A. 1997. Synergistic effect of environmental estrogens: report withdrawn. *Science* 277:462–463.

McNabb, K., J. Borges, and J. Welker. 1994. Jari at 25. *Journal of Forestry* 92(2):21–26.

McNaughton, S. J. 1979. Grassland–herbivore dynamics. In *Serengeti: Dynamics of an Ecosystem*, eds. A. R. E. Sinclair and M. Norton-Griffiths, 46–79. Chicago: Univ. of Chicago Press.

McNaughton, S. J. 1993. Grasses and grazers, science and management. *Ecological Applications* 3:17–20.

Medina, H. V. 1992. *Los Chachi: Supervivencia y Ley Tradicional*. Quito: Abya-Yala.

Mesquita, R. C. G. 1995. The effect of different proportions of canopy opening on the carbon cycle of a central Amazonian secondary forest. Ph.D. dissertation, University of Georgia, Athens.

Mikkelsen, T. R., B. Andersen, and R. B. Jörgensen. 1996. The risk of crop transgene spread. *Nature* 380:31.

Mitchell, G. C. 1986. Vampire bat control in Latin America. In *Ecological Knowledge and Environmental Problem Solving*, 150–164. Washington, DC: National Academy Press.

Moffat, A. S. 1992. Crop scientists break down barriers at Ames meeting. *Science* 257:1347–1348.

Montagnini, F., and F. Sancho. 1994. Above-ground biomass and nutrients in young plantations of four indigenous tree species: implications for site nutrient conservation. *Journal of Sustainable Forestry* 1:115–139.

Montagnini, F., E. Gonzales, C. Porras, and R. Rheingans. 1995a. Mixed and pure forest plantations in the humid neotropics: a comparison of early growth, pest damage and establishment costs. *Commonwealth Forestry Review* 74:306–314.

Montagnini, F., A. Fanzeres, and S. Guimares. 1995b. The potentials of 20 indigenous tree species for soil rehabilitation in the Atlantic forest region of Bahia, Brazil. *Jour. of Applied Ecology* 32:841–856.

Mundt, C. C. 1990. Disease dynamics in agriculture. In *Agroecology*, eds. C. R. Carroll, J. H. Vandermeer, and P. Rosset, 263–299. New York: McGraw-Hill.

Myers, N. 1984. *The Primary Source*. New York: Norton.

National Research Council. 1989. *Alternative Agriculture*. Washington, DC: National Academy Press.

Nations, J. D. 1993. Does conservation condemn the poor to perpetual poverty? A nongovernmental organization perspective. In *Agriculture and Environmental Challenges. Proceedings of the Thirteenth Agricultural Sector Symposium*, eds. J. P. Srivastava and H. Alderman, 245–250. Washington, DC: The World Bank.

Neate, S. M. 1994. Soil and Crop Management Practices that affect root diseases of crops. In *Soil Biota: Management in Sustainable Farming Systems*, eds. C. F. Pankhurst, B. M. Doube, V. V. S. R. Gupta, and P. R. Grace, 96–106. East Melbourne: CSIRO.

New York Times. 1996. Weevil program embitters Texas cotton growers. *National News*, 4 August, 15.

Nilsson, U. 1993. *Competition in Young Stands of Norway Spruce and Scots Pine*. Alnarp: Swedish Univ. of Agricultural Sciences, Southern Swedish Forest Research Centre.

Ninio, J. 1983. *Molecular Approaches to Evolution*. Princeton, NJ: Princeton Univ. Press.

Odum, E. P. 1969. The strategy of ecosystem development. *Science* 164:262–270.

Odum, H. T. 1988. Self-organization, transformity, and information. *Science* 242:1132–1139.

Odum, H. T., and E. C. Odum. 1981. *Energy Basis for Man and Nature*. New York: McGraw-Hill.

Odum, H. T., and R. C. Pinkerton. 1955. Time's speed regulator: the optimum efficiency for maximum power output in physical and biological systems. *American Scientist* 43:331–343.

Olasantan, F. O., H. C. Ezumah, and E. O. Lucas. 1996. Effects of intercropping with maize on the micro-environment, growth and yield of cassava. *Agriculture, Ecosystems and Environment* 57:149–158.

Olson, J. S. 1963. Energy storage and the balance of producers and decomposers in ecological systems. *Ecology* 44:322–332.

Opie, J. 1993. *Ogallala: Water for a Dry Land*. Lincoln: Univ. of Nebraska Press.

Oppert, B., K. J. Kramer, R. W. Beeman, D. Johnson, and W. H. McGaughey. 1997. Proteinase-mediated insect resistance to *Bacillus thuringiensis* toxins. *Journal of Biological Chemistry* 272(38):23473–23476.

Palti, J. 1981. *Cultural Practices and Infectious Crop Diseases*. Berlin: Springer-Verlag.

Palmer, J. R. 1986. Jari: lessons for land managers in the tropics. Paper presented at the International Workshop on Rainforest Regeneration and Management, Guri, Venezuela, 24–28 November, under the auspices of the UNESCO Man and the Biosphere Program.

Paoletti, M. G., and D. Pimentel. 1996. Genetic engineering in agriculture and the environment. *BioScience* 46:665–673.

Park, T. K. 1992. Early trends toward class stratification: chaos, common property, and flood recession agriculture. *American Anthropologist* 94:90–117.

Parsons, J. W. 1985. Organic Farming. In *Soil Organic Matter and Biological Activity*, eds. D. Vaughan and R. E. Malcolm, 423–443. Dordrecht: Junk.

Parton, W. J., D. S. Ojima, and D. S. Schimel. 1996. Models to evaluate soil organic matter storage and dynamics. In *Structure and Organic Matter Storage in Agricultural Soils*, eds. M. R. Carter and B. A. Stewart, 421–448. Boca Raton, FL: Lewis Publishers.

Patten, B. C. 1991. Network ecology: indirect determination of the live-environment relationship in ecosystems. In *Theoretical Studies of Ecosystems: The Network Perspective*, eds. M. Higashi and T. P. Burns, 288–351. Cambridge: Cambridge Univ. Press.

Pedigo, L. P. 1996. *Entomology and Pest Management*. Upper Saddle River, NJ: Prentice-Hall.

Pell, A. N., J. E. Bryant, and L. Anguish. 1994. *Giardia* and *Cryptosporidium parvum*: contagion and containment. In *Liquid Manure Application Systems: Design, Management and Environmental Assessment*, 85–94. Proceedings from a Conference, Rochester, New York. Ithaca, NY: N.E. Agricultural Engineering Service.

Perera, A. H., and R. M. N. Rajapakse. 1991. A baseline study of Kandyan forest gardens of Sri Lanka: structure, composition and utilization. In *Agroforestry: Principles and Practice*, ed. P. G. Jarvis, 269–280. Amsterdam: Elsevier.

Perfecto, I., R. A. Rice, R. Greenberg, and M. E. Van der Voort. 1996. Shade coffee: a disappearing refuge for biodiversity. *BioScience* 46:598–608.

Phelan, P. L., J. F. Mason, and B. R. Stinner. 1995. Soil-fertility management and host preference by European corn borer, *Ostrinia nubilalis* (Hübner), on *Zea mays* L.: A comparison of organic and conventional chemical farming. *Agriculture, Ecosystems and Environment* 56:1–8.

Pimentel, D., and M. Pimentel, eds. 1996. *Food, Energy and Society*. Niwot: Univ. Press of Colorado.

Pimentel, D., L. E. Hurd, A. C. Bellotti, M. J. Forster, I. N. Oka, O. D. Sholes, and R. J. Whitman. 1973. Food production and the energy crisis. *Science* 182:443–449.

Pinard, M. A., F. E. Putz, J. Tay, and T. E. Sullivan. 1995. Creating timber harvest guidelines for a reduced-impact logging project in Malaysia. *Journal of Forestry* 93(10):41–45.

Ponting, C. 1990. Historical perspectives on sustainable development. *Environment* 32(9):4–33.

Posey, D. A. 1982. The keepers of the forest. *Garden* 6:18–24.

Power, A. G. 1990. Cropping systems, insect movement, and the spread of insect-transmitted diseases in crops. In *Agroecology*, ed. S. R. Gliessman, 47–69. New York: Springer-Verlag.

Pretty, J. N. 1995. *Regenerating Agriculture: Policies and Practice for Sustainability and Self-Reliance*. Washington, DC: Joseph Henry Press.

Putnam, A. R., and W. B. Duke. 1974. Biological suppression of weeds: evidence for allelopathy in accessions of cucumber. *Science* 185:370–372.

Raich, J. 1996. Influences of deforestation on the epidemiology of the Chachi of Ecuador. Term Paper, Univ. of Georgia, Athens.

Rasmussen, J., and J. Ascard. 1995. Weed control in organic farming. In *Ecology and Integrated Farming Systems*, eds. D. G. Glen, M. P. Graves, and H. M. Anderson, 49–67. Chichester: Wiley.

Rauber, P. 1996. Poisonberries. *Sierra* 81(4):20–21.

Rawles, K. 1996. Ethical implications of the Gaia thesis. In *Gaia in Action*, ed. P. Bunyard, 308–323. Edinburgh: Floris Books.

Real, L. A. 1996. Sustainability and the ecology of infectious disease. *BioScience* 46:88–97.

Regal, P. 1988. The adaptive potential of genetically engineered organisms in nature. *Trends in Ecology and Evolution* 3:S36–S38 / *Trends in Biotechnology* 6:S36–S38 (combined issue).

Repetto, R., and S. S. Baliga. 1996. *Pesticides and the Immune System: The Public Health Risks*. Washington, DC: World Resources Institute.

Risch, S. J. 1981. Insect herbivore abundance in tropical monocultures and polycultures: an experimental test of two hypotheses. *Ecology* 62:1325–1340.

Rissler, J., and M. Mellon. 1996. *The Ecological Risks of Engineered Crops*. Cambridge: The MIT Press.

Root, R. B. 1973. Organization of a plant-arthropod association in simple and diverse habitats: the fauna of collards (*Brassica oleracea*). *Ecological Monographs* 43:95–124.

Ruehle, J. L., and D. H. Marx. 1979. Fiber, food, fuel and fungal symbionts. *Science* 206:419–206.

Russell, C. E. 1983. Nutrient cycling and productivity of native and plantation forests at Jari Florestal, Para, Brazil. Ph.D. dissertation, Univ. of Georgia, Athens.

Salwasser, H. 1986. Conserving a regional spotted owl population. In *Ecological Knowledge and Environmental Problem Solving*, 227–247. Washington, DC: National Academy Press.

Savory, A. 1988. *Holistic Resource Management*. Washington, DC: Island Press.

Scott, N. M. 1985. Sulphur in Soils and Plants. In *Soil Organic Matter and Biological Activity*, eds. D. Vaughan and R. E. Malcolm, 379–401. Dordrecht: Junk.

Service, R. F. 1996. Pacific Basin gathering in Hawaii fills with chemists. *Science* 271:145–146.

Silcox, C. A., and E. S. Roth. 1995. Pyrethrum for control of pests of agricultural and stored products. In *Pyrethrum Flowers*, eds. J. E. Casida and G. B. Quistad, 287–310. New York: Oxford Univ. Press.

Sioli, H. 1973. Recent human activities in the Brazilian Amazon region, and their ecological effects. In *Tropical Forest Ecosystems in Africa and South America: A Comparative Review*, eds. B. J. Meggars, E. S. Ayensu, and W. D. Duckworth, 321–344. Washington, DC: Smithsonian Institution Press.

Skovgard, H., and P. Päts. 1997. Reduction of stemborer damage by intercropping maize with cowpea. *Agriculture, Ecosystems and Environment* 62:13–19.

Smith, F. E. 1971. *Conservation in the United States, A Documentary History: Land and Water, 1492–1900*. New York: Van Nostrand Reinhold.

Soermarwoto, O. 1977. Nitrogen in tropical agriculture: Indonesia as a case study. *Ambio* 6:162–165.

Soule, J. D., and J. K. Piper. 1992. *Farming in Nature's Image*. Washington, DC: Island Press.

Soulé, M. E. 1991. Conservation: tactics for a constant crisis. *Science* 253:744–750.

Southgate, D., and J. Elegegren. 1995. Development of tropical timber resources by local communities: a case study from the Peruvian Amazon. *Commonwealth Forestry Review* 74:142–146.

Stark, N. M., and C. F. Jordan. 1978. Nutrient retention by the root mat of an Amazonian rain forest. *Ecology* 59:434–437.

Steinhart, C. E., and J. S. Steinhart. 1974. *Energy: Sources, Use, and Role in Human Affairs*. North Scituate, MA: Duxbury Press.

Stokes, B. 1983. Water shortages: the next energy crisis. *The Futurist*, April, 38–47. Cited in Soule, J., D. Carre, and W. Jackson. 1990. Ecological impact of modern agriculture. In *Agroecology*, eds. C. R. Carroll, J. H. Vandermeer, and P. Rosset, 165–188. New York: McGraw-Hill.

Stone, R. 1992. A biopesticidal tree begins to blossom. *Science* 255:1070–1071.

Subler, S., and C. Uhl. 1990. Japanese agroforestry in Amazonia: a case study in Tomé-Açu, Brazil. In *Alternatives to Deforestation: Steps Toward Sustainable Use of the Amazon Rain Forest*, ed. A. B. Anderson, 152–166. New York: Columbia Univ. Press.

Suresh, K. K., and R. S. Vinaya Rai. 1987. Studies on the allelopathic effects of some agroforestry tree crops. *International Tree Crops Journal* 4:109–115.

Swift, M. J., and J. M. Anderson. 1993. Biodiversity and ecosystem function in agricultural systems. In *Biodiversity and Ecosystem Function*, eds. E. D. Schulze and H. A. Mooney, 14–41. Berlin: Springer-Verlag.

Swift, M. J., O. W. Heal, and J. M. Anderson. 1979. *Decomposition in Terrestrial Ecosystems*. Berkeley: Univ. of California Press.

Tabashnik, B. E. 1994. Evolution of resistance to *Bacillus thuringiensis*. *Annual Review Entomology* 39:47–79.

Tabashnik, B. E., Y. Liu, T. Malvar, D. G. Heckel, L. Masson, V. Ballester, B. Granero, J. L. Mensua, and J. Ferre. 1997. Global variation in the genetic and biochemical basis of diamondback moth resistance to *Bacillus thuringiensis*. *Proceedings of the National Academy of Sciences* 94:12780–12785.

Tamm, C. O., H. Holmen, B. Popovic, and G. Wiklander. 1974. Leaching of plant nutrients from soil as a consequence of forestry operations. *Ambio* 3:211–221.

Tan, K. H. 1993. *Principles of Soil Chemistry*. New York: Dekker.

Thurman, E. M., D. A. Goolsby, M. T. Meyer, M. S. Mill, M. L. Pomes, and D. W. Kolpin. 1992. A reconnaissance study of herbicides and their metabolites in surface water of the midwestern United States using immunoassay and gas chromatography/mass spectrometry. *Environmental Science and Technology* 26:2440.

Tilman, D., and J. A. Downing. 1994. Biodiversity and stability in grasslands. *Nature* 367:363–365.

Tilman, D., D. Wedin, and J. Knops. 1996. Productivity and sustainability influenced by biodiversity in grassland ecosystems. *Nature* 379:718–720.

Trenbath, B. R. 1986. Resource use by intercrops. In *Multiple Cropping Systems*, ed. C. A. Francis, 57–81. New York: Macmillan.

Thresh, J. M. 1983. Progress curves of plant virus disease. *Adv. in Applied Biology* 8:1–85.

Union of Concerned Scientists. 1993. Compilation of data from applications to the U.S. Department of Agriculture to field test transgenic crops. Washington, DC. Cited in Rissler, J., and M. Mellon. 1996. *The Ecological Risks of Engineered Crops*, 10. Cambridge: The MIT Press.

Uhl, C., and R. Buschbacher. 1985. A disturbing synergism between cattle ranch burning practices and selective tree harvesting in the Eastern Amazon. *Biotropica* 17:265–268.

Uhl, C., P. Barreto, A. Verissimo, E. Vidal, P. Amaral, A. C. Barros, C. Souza, J. Johns, and J. Gerwing. 1997. Natural resource management in the Brazilian Amazon. *BioScience* 47:160–168.

Vandermeer, J. H. 1990. Intercropping. In *Agroecology*, eds. C. R. Carroll, J. H. Vandermeer, and P. Rosset, 481–516. New York: McGraw-Hill.

Vitousek, P. M., and W. A. Reiners. 1975. Ecosystem succession and nutrient retention: a hypothesis. *BioScience* 25:376–381.

Vitousek, P. M., and P. S. White. 1981. Process studies in succession. In *Forest Succession: Concepts and Application*, eds. D. C. West, H. H. Shugart, and D. B. Botkin, 267–276. New York: Springer-Verlag.

Volobuev, V. R. 1964. *Ecology of Soils*. Jerusalem: Israel Program for Scientific Translations, Ltd.

Wadman, M. 1996. Row over plan to treat plants as "pesticides." *Nature* 382:485.

Walker, L. C. 1984. Thinning the Southern pine plantation: an overview. In *Thinning Southern Pine Plantations Workshop*, 1–19. Washington, DC: American Pulpwood Assoc. Inc.

Wargo, J. 1996. *Our Children's Toxic Legacy*. New Haven: Yale Univ. Press.

Weiner, J. 1990. Plant population ecology in agriculture. In *Agroecology*, eds. C. R. Carroll, J. H. Vandermeer, and P. Rosset, 235–262. New York: McGraw-Hill.

White, P. S., and R. D. White. 1996. Old-growth oak and oak-hickory forests. In *Eastern Old Growth Forests*, ed. M. B. Davis, 178–198. Washington, DC: Island Press.

Willmer, P. G., and G. N. Stone. 1997. How aggressive ant-guards assist seed-set in *Acacia* flowers. *Nature* 388:165–167.

Woomer, P. L., and M. J. Swift. 1994. *The Biology and Fertility of Tropical Soils*. Nairobi, Kenya: Tropical Soil Biology and Fertility Programme.

Young, A. 1989. *Agroforestry for Soil Conservation*. Wallingford, UK: CAB International.

Zachary, G. P. 1996. Global food supplies: tighter but adequate. *The Wall Street Journal*, 15 July, 228:1.

SUBJECT INDEX

A

Agroecology, 27
Agroforestry, 27, 28, 30, 65, 95
Allelopathy, 39
Alley cropping, 43, 65
Alternative agriculture, 27, 30

B

Bacillus thuringiensis, 17, 18
Biological control, of pests and diseases, 74–80
Boll weevil, 11
Bovine growth hormone, 14

C

Clean Air Act, 11
Co-evolution, 106, 109

Competition, 35, 36
Complementarity, 40, 47, 49
   ecological, 41, 45
   economic, 46
Conservation tillage, 63, 64
Cotton, spraying of, 11
Crop
   polycultures, 69–72, 76, 77
   rotations, 63

D

Decomposition, 100, 102
Double cropping, 42
DDT, 10
Dioxin, 12
Disease
   crop, 69, 71–74, 80–83
   in polycultures, 72
   resistance, 72, 73
Disturbance, and succession, 85–87
Diversity, functions resulting from, 55, 56

E

Ecological management
  of pests, 76
Economic
  analysis for pest control, 80
  considerations, 66, 136
  feasibility, 28, 96
Ecosystem
  efficiency of energy use, 110
  evolution, 105–107
  productivity, 101, 102
Energy
  analysis, 110
  flow through soil, 52, 53, 56, 58
  and power output, 103–105
  subsidies, 61
Environmental Protection Agency,
  11, 18
Extractivism, 116–117

F

Facilitation, 33, 40–41, 49
Fallowing, 39, 44, 49, 95
Forest
  management, 92
  plantations, 35, 45–49, 66
Forestry
  sustainable, 65
  tropical, 116–122, 128, 129
Forest Service, 7

G

Gaia, 107, 108
Genetically altered organisms, 79
Genetic diversity, 73
Genetic engineering, 17–19
  risks of, 18, 19
Genetics, 14–16
Grasslands
  management of, 93
Grazing systems, 93, 122–125,
  130, 131
Green manure, 44, 45

H

Herbicides, 12, 13
  resistance to, 19
Holistic resource management, 27
Hormones, 14

I

Indigenous agriculture, 94, 95
Industrial Revolution, 5, 9
Insecticides, natural, 75, 76
Insects, pests, 69–71
Integrated pest management,
  73–80
Interactions in Nature, 28, 31–34
  during succession, 85
  indirect, 33
  pests and disease, 69
  plant/plant, 35
  soil, 51

Intercropping, 41–43
Irrigation, effects of scale, 25

J

Jari, 119–121

L

Land Institute, 96
Logging
 creaming, 121
 high grading, 121
 reduced impact, 128
Low-input agriculture, 27
Lyme disease, 81, 82

M

Malaria, 82
Malthus, Thomas, 2
Manure, environmental problems, 22, 23
Mass trapping of insects, 79
Mating disruption, for insect control, 79
Methyl bromide, 11
Microclimate, in soil, 57
Monitoring, for disease and pests, 78

Montreal Protocol, 11
Morrill Act, 7
Mulch, 37, 38
Mutualism, 32
Mycorrhizae, 36, 55

N

Natural enemy hypothesis, 69
Nurse species, 48
Nutrient cycling, 52–54
 efficiency of, 100, 102

O

Onchocercosis, 82
Optimal spacing, 36
Organic agriculture, 27, 62, 63
Overyielding, 40

P

Permaculture, 27
Pest management, 69
Pesticides, 9–12
 resistance to, 18
Pests, insect and disease, 69, 71
Photoactive dyes, 80
Plantations, 46–48, 119–121

Potato famine fungus, 15
Power output, 103
  and succession, 105
  of systems, 103, 104
Prairies, 93, 96
Precautionary principle, 3
Primary productivity, 100, 101

R

Range management, 122–125, 130, 131
Resource concentration hypothesis, 70
Resource management
  case studies, 113
  principle of, 111
Risk, reduction of, 43
Rotations, of crops, 45, 56

S

Sahel, 114–116
Scale of farm and forest management, effects of, 21
Selection cutting, 117
Serengeti, 122–125
Shifting cultivation, 39, 44
Sierra Club, 8
Society of American Foresters, 7
Soil Conservation Service, 8

Soil
  disease suppression, 56, 57, 60
  effects of cultivation, 58, 59
  microclimate, 57
  organic matter, 52–54
  organisms, 51
  structure, 56, 57, 60
Spacing, in plantations, 36, 37, 45
Species Diversity, 100, 103,
  loss of, 25, 26
Standing stock, 100, 101
Sterile insects technique, 79, 80
Strip cutting, 118, 119
Structural diversification, 48, 49, 55, 56
Subsidies, 61, 66,
Succession, 85
  and forest management, 91, 92
  and grazing systems, 93
  and nitrogen, 90, 91
  and phosphorus, 90, 91
  in Southeastern U.S., 88–90
Successional mimics in agriculture, 96
Sustainable resource management, 1, 27, 29
  case studies of, 125–131
Symbiosis, 32

T

Thermodynamics, first law of, 16
Tillage
  conservation tillage, 14, 63, 64
  minimum tillage, 14, 64
  no till, 13, 14, 64
Tomé Açu, Brazil, 125–128

Weeds
  control of, 37–39
  traits of, 36

Author Index

A

Addiss, D. G., 22
Ae, N. J., 41
Alegre, J. C., 44
Alstad, D. N., 18
Altieri, M. A., 38, 69, 70, 76, 77
Amaral, P., 129
Andersen, B., 19
Anderson, J. M., 55
Andow, D., 18, 77
Anguish, L., 23
Annis, P. C., 11
Aracil, E., 74
Arihara, J., 41
Arnold, S. F., 10
Ascard, J., 38
Athari, S., 92
Atta-Krah, A. N., 43, 65
Avery, D. T., 2, 10

B

Ba, T. A., 115
Baliga, S. S., 9
Barlow, H. S., 79
Barreto, P., 129
Barrett, S. A., 82
Barrett, G. W., 60
Barros, A. C., 129
Baskerville G. L., 49
Batmanian, G., 47
Baumgartner, A., 58
Beare, D. A., 54, 60
Beeman, R. W., 117
Bellotti, A. C., 61
Benedict, F., 41
Benenson, A. S., 82
Berish, C., 41
Bhat, B. K., 75
Blair, K. A., 22
Bohlen, P. J., 60
Bolick, M. R., 13
Borges, J., 121
Boucher, D. H., 32
Brady, N. C., 54
Browder, J. O., 117
Brown, B., 41
Bryant, J. E., 23
Bucher, E. H., 117
Buckley, J., 10
Burkart, M. R., 13
Buschbacher, R., 122

C

Carroll, C. R., 74
Chapin, F. S., 103
Chatterton, B., 45
Chatterton, L., 45
Cheshire, M. V., 57
Child, R. D., 130
Chou, C. H., 39
Clark, J. B., 18
Colborn, T., 9, 11, 12, 14
Coleman, D. C., 53, 54, 60
Collins, B. M., 10
Cook, R. J., 56
Costanza, R., 66
Crossley, D. A., 53, 54, 60
Crousse B., 115
Culotta, E., 78
Cussans, G. W., 38

D

Davis, J. P., 22
DeBlieu, J., 82
DelFosse, E., 74
del Moral, R., 39
Desowitz, R. S., 83
DeWitt, J. C., 12
Dodson, C. H., 82
Doebley, J. F., 16
Downing, J. A., 103
Dubois, J. C. L., 116
Duke, W. B., 40
Dumanoski, D., 9, 11, 12, 14

E

Elegegren J., 119
Erdle, T. A., 49
Evans, J., 47
Ewel, J., 41
Ezuhah, H. C., 42

F

Fanzeres, A., 55
Farnworth E. G., 48
Federici, B. A., 79
Ferguson, B. K., 25
Ferrell, W. K., 65
Fineblum, W. L., 16
Folke, C., 66
Forster, M. J., 61
Fox, J. E. D., 117
Fox, K. R., 22
Francis, C. A., 33, 42, 43
Franklin, J. F., 65
Fry, W. E., 15
Fulhage, C., 22

G

Gajaseni, J., 46, 47
Gentry, A., 82
Gershuny, G., 62
Gerwing, J. J., 66, 129
Gillette, D., 22, 23

Glaser D. L., 70
Gliessman, S. R., 39
Gonzalez, E., 47
Gonzalez Andujar, J. L., 39
Goodwin S. B., 15
Goolsby, D. A., 13
Gordon, J. C., 7, 24
Grace, P. R., 58
Gradus, M. S., 22
Greenberg, R., 76
Guedes, R., 41
Guilette, L. J., 10
Guimares, S., 55
Gupta, V. V. S. R., 54
Guzmán, R. M., 16

Halstead, P., 115
Hammer, M., 66
Harmon, M. E., 65
Harrison, H. L., 10
Hart, R. D., 95
Hartshorn, G. S., 118
Hasan, S., 74
Hasse, V., 70
Hauser, S., 43, 65
Heady, H. F., 130
Heal, O. W., 55
Heitz, J. R., 80
Hendrix, P. F., 54, 60
Henshel, D. S., 12
Hobbs, R. J., 103
Hoffman, M. P., 78
Holmen, H., 24

Homewood, K. M., 123, 124, 125
Hooper, D. U., 103
Hornick, J. R., 133
Hoxie, N. J., 22
Hu, S., 54, 60
Hughes, G., 39
Huo, Y., 42
Hurd, L. E., 61

Iltis, H. H., 16
International Society of Tropical Forestry, 121

Jannson, A., 66
Janzen, D. H., 106
Johansen, C., 41
Johns, J. S., 66, 129
Johnson, D., 19
Jordan, C.F., 13, 24, 36, 38, 43, 44, 46, 47, 48, 53, 54, 58, 66, 71, 85, 101, 112, 119, 126
Jörgensen, R. B., 19

Kadir, A. A. S. A., 79
Kaiser, J., 9
Kang, B. T., 43, 65
Kazmierczak, J. J., 22
Keeler, K. H., 13

Keever, C., 89
Kirchner, M., 58
Kjaergaard, T., 45
Kloepper, J. W., 74
Klotz, D. M., 10
Knops, J., 103
Kolpin, D. W., 13
Kramer H., 92
Kramer, K. J., 19
Krimsky, S., 14
Kropff, M. J., 37

L

Ladd, J. N., 58
Lavelle, P., 107
Lawton, J. H., 103
Lewis, D. H., 33
Lewis, R. C., 74
Liebman, M., 69, 76
Line, L., 12
Litsinger J. A., 70
Longley, M., 12
Lotz, L. A. P., 37
Loucks, O. L., 10
Lovelock, J., 108
Lowe, R. G., 118
Lucas, E. O., 42
Luck, R. F., 74

M

Macilwain, C., 18
MacKenzie, W. R., 22
MacNeil, C. R., 78

Maddox J. V., 79
Margulis, L., 108
Martin, D. L., 62
Martin, J. W., 12
Marx, D. H., 55
Mason, J. F., 57
Matta-Machado, R. P., 38
McEvoy, P. B., 74, 79
McGaughey, W. H., 18, 19
McGuire, R. T., 23
McIntosh, R. J., 114
McLachlan, J. A., 10
McNabb, K., 121
McNaughton, S. J., 6, 122, 125
Medina, H. V., 82
Mellon, M., 17, 18
Mesquita, R. C. G., 48
Meyer, D. A., 13
Mikkelsen, T. R., 19
Mill, M. T., 13
Mishanec, M. S., 78
Mitchell, G. C., 75
Mitchell, J. W., 10
Moffat, A. S., 21
Montagnini, F., 47, 55
Muller C. H., 39
Mundt, C. C., 72
Myers, J. P., 9, 11, 12, 14
Myers, N., 122

N

National Research Council, 12, 13, 15, 61, 63
Nations, J. D., 117
Neate, S. M., 60

New York Times, 11
Nilsson, U., 45
Ninio, J., 18

O

Odum, E. C., 53
Odum, E. P., 87, 105
Odum, H. T., 53, 104, 109
Ojima, D. S., 51
Oka, I. N., 61
Okada, K., 41
Olasantan, F. O., 42
Olson, J. S., 100
Opie, J., 25
Oppert, B., 19
Orfanedes, M. S., 78
O'Shea, J., 115

P

Palti, J., 71
Palmer, J. R., 120
Paoletti, M. G., 19
Park, T. K., 114
Parkhurst, D. F., 10
Parsons, J. W., 57
Parton, W. J., 51
Päts P., 77
Patten, B. C., 33, 106
Pazy, B., 16
Pedigo, L. P., 76, 80
Pell, A. N., 23
Perera, A. H., 42
Perfecto, I., 76

Peterson, D. E., 22
Petzoldt, C. H., 78
Phelan, P. L., 57
Pimentel, D., 19, 61, 110
Pimentel, M., 110
Pinard, M. A., 128
Pinkerton R. C., 104, 109
Piper J. K., 96
Pomes, M. L., 13
Ponting, C., 20
Popovic, B., 24
Porras, C., 47
Posey, D. A., 94
Power, A. G., 71
Proctor, M. E., 22
Putnam, A. R., 40
Putz, F. E., 128

R

Rai, R. S. V., 39
Raich, J., 83
Rajapakse, R. M. N., 42
Rao, M. R., 44
Rasmussen, J., 38
Rauber, P., 11
Rausher, M. D., 16
Rawles, K., 108
Real, L. A., 73, 81
Regal, P., 18
Reiners, W. A., 101
Repetto, R., 9
Reynolds, L., 43, 65
Rheingans, R., 47
Rice, R. A., 76
Risch, S. J., 70, 74, 76

Rissler, J., 17, 18
Rodenhouse, N., 60
Rogers W. A., 123, 124, 125
Root, R. B., 70
Rose, J. B., 22
Roth, E. S., 75
Ruehle, J. L., 55
Russell, C. E., 119, 120

Sala, O. E., 103
Salwasser, H., 24
Sancho, F., 47
Sanginga, N., 43, 65
Savory, A., 93, 131
Schimel, D. S., 51
Schmidt L. L., 70, 77
Scott, N. M., 66
Service, R. F., 19
Sholes, O. D., 61
Silcox, C. A., 75
Sioli, H., 125
Skjemstad, J. O., 58
Skovgard, H., 77
Smith, F. E., 7
Soermarwoto, O., 63
Sotherton N. W., 12
Soule, J. D., 96
Soulé, M. E., 26
Southgate, D., 119
Souza, C., 129
Stark, N. M., 54
Steinhart, C. E., 110
Steinhart J. S., 110
Stinner, B. R., 57

Stokes, B., 25
Stone, G. N., 74
Stone, R., 75, 76
Subler, S., 128
Sullivan, T. E., 128
Suresh, K. K., 39
Swift, M. J., 55, 56

Tabashnik, B. E., 18, 19
Tamm, C. O., 24
Tan, K. H., 58
Tay, J., 128
Thresh, J. M., 72
Thurman, E. M., 13
Tilman, D., 103
Tracy, D. G., 10
Turner, C. E., 13
Trenbath, B. R., 40

U

Uhl, C., 122, 128, 129
Union of Concerned Scientists, 17

V

Vandermeer, J. H., 33
Van der Voort, M. E., 76
Vanlauwe, B., 43, 65
Van Vliet, P. C. J., 54, 60
Verissimo, A., 129
Vidal, E., 66, 129

Vinaya Rai, R. S., 39
Vitousek, P. M., 91, 101
Volobuev, V. R., 108
Vonier, L. J., 10

Wadman, M., 17
Walker, B. H., 103
Walker, L. C., 45
Wallinga, J., 37
Wargo, J., 10
Waterford, C. J., 11
Watts, D. G., 10
Wedin, D., 103
Weiner, J., 39
Welker, J., 121
Whalon, M. E., 18, 19
White, P. S., 88, 91
White. R.D., 88

Whitman, R. J., 61
Whitmore, J. L., 133
Wicklander, G., 24
Willmer, P. G., 74
Woomer, P. L., 56
Wrubel, R. P., 14

Yannacone, V. J., 10
Young, A., 65
Young, D. H., 78
Yoshihara T., 41

Zachary, G. P., 136
Zerbe, J. I., 133